W0036837

Principles of Electronic Instrumentation

Principles of Electronic Instrumentation

Contributors

El-Hassane Aglzim, and Amar Rouane et al.

AURIS Reference

www.aurisreference.com

Principles of Electronic Instrumentation

Contributors: El-Hassane Aglzim ,and Amar Rouane et al.

Published by Auris Reference Limited
www.aurisreference.com

United Kingdom

Copyright 2016
Printed in 2017 for Sale in the Indian Subcontinent

The information in this book has been obtained from highly regarded resources. The copyrights for individual articles remain with the authors, as indicated. All chapters are distributed under the terms of the Creative Commons Attribution License, which permit unrestricted use, distribution, and reproduction in any medium, provided the original author and source are credited.

Notice

Contributors, whose names have been given on the book cover, are not associated with the Publisher. The editors and the Publisher have attempted to trace the copyright holders of all material reproduced in this publication and apologise to copyright holders if permission has not been obtained. If any copyright holder has not been acknowledged, please write to us so we may rectify.

Reasonable efforts have been made to publish reliable data. The views articulated in the chapters are those of the individual contributors, and not necessarily those of the editors or the Publisher. Editors and/or the Publisher are not responsible for the accuracy of the information in the published chapters or consequences from their use. The Publisher accepts no responsibility for any damage or grievance to individual(s) or property arising out of the use of any material(s), instruction(s), methods or thoughts in the book.

Principles of Electronic Instrumentation

ISBN: 978-1-78154-925-4

British Library Cataloguing in Publication Data
A CIP record for this book is available from the British Library

Printed in the United Kingdom

Exclusively distributed by CBS Publishers & Distributors Pvt. Ltd.

Sales & Distribution Rights only for India, Pakistan, Bangladesh, Sri Lanka, Nepal and Bhutan. This book is not to be sold outside these territories.

Contents

List of Abbreviations .... *vii*

List of Contributors .... *ix*

Preface .... *xiii*

Chapter 1   **An Electronic Measurement Instrumentation of the Impedance of a Loaded Fuel Cell or Battery** 1

El-Hassane Aglzim, Amar Rouane and Reddad El-Moznine

Chapter 2   **RICE: A Reliable and Efficient Remote Instrumentation Collaboration Environment** 19

Prasad Calyam, Abdul Kalash, Ramya Gopalan, Sowmya Gopalan, and Ashok Krishnamurthy

Chapter 3   **Efficacy of Electronic Foramen Locators in Controlling Root Canal Working Length during Rotary Instrumentation** 55

Lorena Arruda Parente, Martin D. Levin, Rodrigo Ricci Vivan, Ricardo Affonso Bernardes, Marco Antonio Hungaro Duarte, Bruno Carvalho de Vasconcelos

Chapter 4   **The Effect of Instrumental Timbre on Interval Discrimination** 65

Jean Mary Zarate, Caroline R. Ritson, David Poeppel

Chapter 5   **Open-Source Hardware Is a Low-Cost Alternative for Scientific Instrumentation and Research** 85

Daniel K. Fisher, Peter J. Gould

Chapter 6   **Applying Software Engineering Methodology for Designing Biomedical Software Devoted to Electronic Instrumentation** 111

Gilsa Aparecida de Lima Machado, Patricia Mara Danella Zacaro, Alderico Rodrigues de Paula Junior and Marcelo Lopes de Oliveira e Souza

Chapter 7   **Instrumentation for Ferromagnetic Resonance Spectrometer** 137

Chi-Kuen Lo

Chapter 8   **A Novel Instrumentation Circuit for Electrochemical Measurements** 157

Li-Te Yin, Hung-Yu Wang, Yang-Chiuan Lin and Wen-Chung Huang

Chapter 9   Evaluation of Three Instrumentation Techniques at the Precision
             of Apical Stop and Apical Sealing of Obturation 171

             Özgür Genç, Tayfun Alaçam, Guven Kayaoglu

Chapter 10   True Unipolar ECG Machine for Wilson Central Terminal
             Measurements... 183

             Gaetano D. Gargiulo

Chapter 11   Improvement of EEG Signal Acquisition: An Electrical
             Aspect for State of the Art of Front End 199

             Ali Bulent Usakli

Chapter 12   Application Concept of Zero Method Measurement in
             Microwave Radiometers.. 215

             Alexander V. Filatov

Chapter 13   Monitoring Instrumentation in Underground Structures.................. 235

             Alireza Maghsoudi, Behzad Kalantari

Chapter 14   The Fundamental Operating Principles of Electronic Root
             Canal Length Measurement Devices 251

             M. H. Nekoofar, M. M. Ghandi, S. J. Hayes & P. M. H. Dummer

             Citations 281

             Index.. 285

List of Abbreviations

AC	Alternating current
ADC	Analog digital converter
A/D	Analog-to-digital
AAR	Auto-reverse mode
BCI	Brain Computer Interface
CDJ	Cemento-dentinal junction
CAMM	Center for Accelerated Maturation of Materials
CMRR	Common Mode Rejection Ratio
CEI	Computerized Electronic Instrumentation
CPT	Cone penetration Tests
CPW	Co-planar waveguide
DFD	Data Flow Diagram
DUT	Device under test
DC	Direct current
DC	Directional coupler
ESI	Electrical Source Imaging
EIS	Electrochemical Impedance Spectroscopy
EEG	Electroencephalogram
EMI	Electromagnetic interference
EFLs	Electronic foramen locators
ERCLMD	Electronic root canal length measurement device'
ESD	Electrostatic discharge
EDS	Energy dispersive spectroscopy
ERD	Entity and Relationship Diagram
ED	Event detected
EDZ	Excavation Damage Zone
EGFET	Extended gate field effect transistor
FA	False alarm
GB	Gigabyte
GUI	Graphical user interface
IRT	Infrared thermometer
IO	Input/output
IA	Instrumentation amplifier
IDE	Integrated Development Environment
ISO	International Standard Organization
ISFET	Ion-sensitive field effect transistor
KB	Kilobytes
L.I.E.N	Laboratory of Electronic Instrumentation of Nancy
LA	Left arm
LL	Left leg
LEDs	Light-emitting diodes

LOD	Limit-of-detection
LOQ	Limit-of-quantification
LCM	Liquid crystal display module
MA	Modulation amplitude
MF	Modulation frequency
NCMIR	National Center for Microscopy and Imaging Research
NYU	New York University
NG	Noise generator
OSC	Ohio Supercomputer Center's
PB	Phosphate buffer
PMT	Pressure meter tests
PCB	Printed circuit board
QoE	Quality of experience
RTC	Real-time clock
RE	Reference electrode
RICE	Remote Instrumentation Collaboration Environment
RA	Right arm
RL	Right leg
RTT	Round-trip delay
SNA	Scalar network analyzer
SRT	Seismic refraction tomography
SWC	Shorted waveguide cavity
SNR	Signal to Noise Ratio
SE	Software Engineering
SPT	Standard Penetration Test
TC	Time constant
TE	Trigger elevated
UHVEM	Ultra High Voltage Electron Microscopy
UML	Unified Modelling Language
VNA	Vector network analyzer
VNC	Virtual network computing
WCT	Wilson Central Terminal
WE	Working electrode

List of Contributors

El-Hassane Aglzim
Laboratoire d'Instrumentation Electronique de Nancy (L.I.E.N.), Nancy Université, Boulevard des Aiguillettes, BP239 - 54506 Vandoeuvre les Nancy, France

Amar Rouane
Laboratoire d'Instrumentation Electronique de Nancy (L.I.E.N.), Nancy Université, Boulevard des Aiguillettes, BP239 - 54506 Vandoeuvre les Nancy, France

Reddad El-Moznine
Physical Laboratory of the Condensed Matter (L.P.M.C.), Université Chouaib Doukkali, Route Ben Maachou, BP20 - 24000 El Jadida, Morocco

Prasad Calyam
Cyberinfrastructure and Software Development Group, Ohio Supercomputer Center, 1224 Kinnear Road, Columbus, OH 43212, USA

Abdul Kalash
Cyberinfrastructure and Software Development Group, Ohio Supercomputer Center, 1224 Kinnear Road, Columbus, OH 43212, USA

Ramya Gopalan
Cyberinfrastructure and Software Development Group, Ohio Supercomputer Center, 1224 Kinnear Road, Columbus, OH 43212, USA

Sowmya Gopalan
Cyberinfrastructure and Software Development Group, Ohio Supercomputer Center, 1224 Kinnear Road, Columbus, OH 43212, USA

Ashok Krishnamurthy
Cyberinfrastructure and Software Development Group, Ohio Supercomputer Center, 1224 Kinnear Road, Columbus, OH 43212, USA

Lorena Arruda Parente
School of Dental Medicine of Sobral, UFC - Universidade Federal do Ceará, Sobral, CE, Brazil

Martin D. Levin
Department of Endodontics, School of Dental Medicine, University of Pennsylvania, Philadelphia, PA, USA

Rodrigo Ricci Vivan
Department of Dentistry, Endodontics and Dental Materials, Bauru Dental School, USP - Universidade de São Paulo, Bauru, SP,Brazil

Ricardo Affonso Bernardes
ABO - Brazilian Dental Association, Taguatinga, DF, Brazil

Marco Antonio Hungaro Duarte
Department of Dentistry, Endodontics and Dental Materials, Bauru Dental School, USP - Universidade de São Paulo, Bauru, SP,Brazil

Bruno Carvalho de Vasconcelos
School of Dental Medicine of Sobral, UFC - Universidade Federal do Ceará, Sobral, CE, Brazil

Jean Mary Zarate
Department of Psychology, New York University, New York, New York, United States of America

Caroline R. Ritson
Department of Psychology, New York University, New York, New York, United States of America

David Poeppel
Department of Psychology, New York University, New York, New York, United States of America

Daniel K. Fisher
USDA Agricultural Research Service, Stoneville, USA

Peter J. Gould
US Forest Service, Pacific Northwest Research Station, Olympia, USA

Gilsa Aparecida de Lima Machado
Research and Development Institute (IPD) Course of Biomedical Engineering, University of Vale do Paraiba (UNIVAP),Av. Shishima Hifumi, 2911, 12244-000, S. Jose dos Campos, SP, Brazil

Patricia Mara Danella Zacaro
Research and Development Institute (IPD) Course of Biomedical Engineering, University of Vale do Paraiba (UNIVAP),Av. Shishima Hifumi, 2911, 12244-000, S. Jose dos Campos, SP, Brazil

Alderico Rodrigues de Paula Junior
Research and Development Institute (IPD) Course of Biomedical Engineering, University of Vale do Paraiba (UNIVAP),Av. Shishima Hifumi, 2911, 12244-000, S. Jose dos Campos, SP, Brazil

Marcelo Lopes de Oliveira e Souza
Course of Space Engineering and Technology (ETE),Division of Space Mechanics and Control (DMC),
National Institute for Space Research (INPE),Av. dos Astronautas, 1758, 12227-010, S. Jose dos Campos, SP, Brazil

Chi-Kuen Lo
Department of Physics, National Taiwan Normal University, Taipei, Taiwan

Li-Te Yin
Department of Optometry, Chung Hwa University of Medical Technology, Tainan 717, Taiwan

Hung-Yu Wang
Department of Electronic Engineering, National Kaohsiung University of Applied Science, Kaohsiung 807, Taiwan

Yang-Chiuan Lin
Department of Electronic Engineering, National Kaohsiung University of Applied Science, Kaohsiung 807, Taiwan

Wen-Chung Huang
General Education Center, Chung Hwa University of Medical Technology, Tainan 717, Taiwan

Özgür Genç
DDS, PhD, Department of Restorative Dentistry and Endodontics, Faculty of Dentistry, Yüzüncü Yıl University, Van, Turkey

Tayfun Alaçam
DDS, PhD, Department of Restorative Dentistry and Endodontics, Faculty of Dentistry, Gazi University, Ankara, Turkey

Guven Kayaoglu
DDS, PhD, Department of Restorative Dentistry and Endodontics, Faculty of Dentistry, Gazi University, Ankara, Turkey

Gaetano D. Gargiulo
The MARCS Institute, University of Western Sydney, Kingswood, NSW 2747, Australia

Ali Bulent Usakli
Department of Technical Sciences, The NCO Academy, 10100 Balikesir, Turkey

Alexander V. Filatov
Tomsk State University of Control Systems and Radio Engineering, Tomsk, Russia

Alireza Maghsoudi
Department of Civil Engineering, University of Hormozgan, Bandar Abbas, Iran

Behzad Kalantari
Department of Civil Engineering, University of Hormozgan, Bandar Abbas, Iran

M. H. Nekoofar
Department of Endodontics, Faculty of Dentistry, Tehran University of Medical Science, Tehran, Iran
Division of Adult Dental Health, School of Dentistry, Cardiff University, Cardiff, UK; and

M. M. Ghandi
Department of Electronic Systems Engineering, University of Essex, Colchester, UK

S. J. Hayes
Division of Adult Dental Health, School of Dentistry, Cardiff University, Cardiff, UK; and

P. M. H. Dummer
Division of Adult Dental Health, School of Dentistry, Cardiff University, Cardiff, UK; and

Preface

Instrumentation is the development or use of measuring instruments for observation, monitoring or control. An instrument is a device that measures a physical quantity, such as flow, temperature, level, distance, angle, or pressure. The text *Principles of Electronic Instrumentation* discusses how to design, select and operate conventional, virtual, and network-based electronic instruments. In first chapter, we present an inexpensive electronic measurement instrumentation developed in laboratory, to measure and plot the impedance of a loaded fuel cell or battery. In second chapter, we model and characterize the complex interplay between the user control behavior and video image transfer performance in remote instrumentation sessions. Third chapter evaluates the efficacy of electronic foramen locators (EFLs) to control root canal working length during rotary instrumentation and to assess possible reliability variations of different working lengths. The effect of instrumental timbre on interval discrimination has been investigated in fourth chapter. The objective of fifth chapter is to introduce researchers and practitioners to potential applications of the open source Arduino platform for implementation in research and monitoring applications. Sixth chapter reports how the software engineering (SE) approach is applied to design and develop biomedical software, which is part of a computerized electronic instrumentation (CEI). Seventh chapter focuses on instrumentation for ferromagnetic resonance spectrometer. In eighth chapter, a novel signal processing circuit which can be used for the measurement of H^+ ion and urea concentration has been presented. The aim of ninth chapter is to investigate the ability of two NiTi rotary apical preparation techniques used with an electronic apex locator-integrated endodontic motor and a manual technique to create an apical stop at a predetermined level in teeth with disrupted apical constriction. Tenth chapter discusses about true unipolar ECG machine for Wilson central terminal measurements. The aim of eleventh chapter is to present some practical state-of-the-art considerations in acquiring satisfactory signals for electroencephalographic signal acquisition. Twelfth chapter examines microwave radiometer functioning algorithm with synchronously using of the two types of pulse modulation: amplitude pulse modulation and pulse-width modulation. Thirteenth chapter presents the features of sophisticated instrumentation available today for geotechnical monitoring. The aim of last chapter is to clarify the fundamental operating principles of the different types of electronic systems that claim to measure canal length.

Chapter 1

AN ELECTRONIC MEASUREMENT INSTRUMENTATION OF THE IMPEDANCE OF A LOADED FUEL CELL OR BATTERY

El-Hassane Aglzim[1], Amar Rouane[1] and Reddad El-Moznine[2]

[1]Laboratoire d'Instrumentation Electronique de Nancy (L.I.E.N.), Nancy Université, Boulevard des Aiguillettes, BP239 - 54506 Vandoeuvre les Nancy, France

[2]Physical Laboratory of the Condensed Matter (L.P.M.C.), Université Chouaib Doukkali, Route Ben Maachou, BP20 - 24000 El Jadida, Morocco

ABSTRACT

In this paper we present an inexpensive electronic measurement instrumentation developed in our laboratory, to measure and plot the impedance of a loaded fuel cell or battery. Impedance measurements were taken by using the load modulation method. This instrumentation has been developed around a VXI system stand which controls electronic cards. Software under Hpvee® was developed for automatic measurements and the layout of the impedance of the fuel cell on load. The measurement environment, like the ambient temperature, the fuel cell temperature, the level of the hydrogen, etc..., were taken with several sensors that enable us to control the measurement. To filter the noise and the influence of the 50Hz, we have implemented a synchronous detection which filters in a very narrow way around the useful signal. The theoretical result obtained by a simulation under Pspice® of the method used consolidates the choice of this method and the possibility of obtaining correct and exploitable results. The experimental results are preliminary results on a 12V vehicle battery, having an inrush current of 330A and a capacity of 40Ah (impedance measurements on a fuel cell are in progress, and will be the subject of a forthcoming paper). The results were plotted at various nominal voltages of the battery (12.7V, 10V, 8V and 5V) and with two imposed currents (0.6A and 4A). The Nyquist diagram resulting from the experimental data enable us to show an influence of the load of the battery on its internal impedance.

The similitude in the graph form and in order of magnitude of the values obtained (both theoretical and practical) enables us to validate our electronic measurement instrumentation. One of the future uses for this instrumentation is to integrate it with several control sensors, on a vehicle as an embedded system to monitor the degradation of fuel cell membranes.

INTRODUCTION

The challengers in energy and climatic conditions are now very well established. The use of hydrogen in fuel cell is an essential vector, and has been the focus of intensive study in recent years as promising alternative energy sources. Thus, various studies have been carried out in the electronic-physic domain of these kinds of generator [1][2]. Impedance measurement is a powerful technique, which can provide useful information on the electro-chemical systems in a real and very short time [3]. This technique can be considered as a good tool to determine the state of charge of batteries or fuel cells.

To bring a solution to the optimization of the powers of the fuel cells, we invested within the Laboratory of Electronic Instrumentation of Nancy (L.I.E.N) in the development and the realization of a system for the impedance measurement of a fuel cell on load using electronic cards and sensors. Most of impedance measurements on fuel cells are made without load. Impedance measurement on a battery or fuel cell with load is very important in order to examine the influence of this latter, on its performance. For these reasons, we are interested in the development and the realization of a system in order to achieve impedance measurement of a battery or fuel cell on load, in our laboratory. This system is based on the electrochemical impedance spectroscopy (EIS) method for measuring and plotting the diagram Nyquist of battery or fuel cell impedance. The analysis and the shape of the diagram can provide information about the state of charge of the sample under test.

THEORETICAL CONSIDERATION

Method

A simulation test under Pspice® was developed in order to validate the choice of our method and its ability to provide some accurate and exploitable results.

The electrical model to represent the different components is given in Figure 1. The variable load is represented by a MOSFET, and the internal impedance of the fuel cell is represented by Randle›s circuit (R_{ohm} = 10mΩ, R_{act} = 90mΩ, and C_{dc} = 300μF). These values correspond to what one can expect for a PEM fuel cell. These estimated values are based on the work carried out by Noponen [7], Wagner [8] and Brunetto and al. [9]. The later other did measurement on PEMFC with the same dimensions as our fuel cell on which the experimental measurements are expected to be done. They found that the real part is ranging from 3mΩ and 200mΩ and the imaginary part is close to 15mΩ.

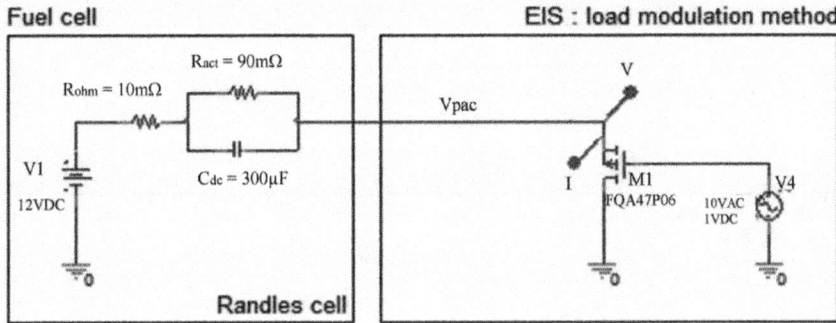

Figure 1: Schematic diagram of the electrical method used for simulation.

Results

The frequency range used in this simulation used here is the same for the experimental measurement. It is ranging from some mHz to 10 kHz. Figure 2 shows the result of this simulation in the Nyquist graph, where the imaginary part (Z›) versus the real part (Z›) of the complex impedance is plotted. A perfect semi-circle is obtained. Also as it can be seen, the negative sign before the imaginary part (Z›). The results of this simulation allowed us to make a comparison with experimental results in order to validate this method and to examine its feasibility.

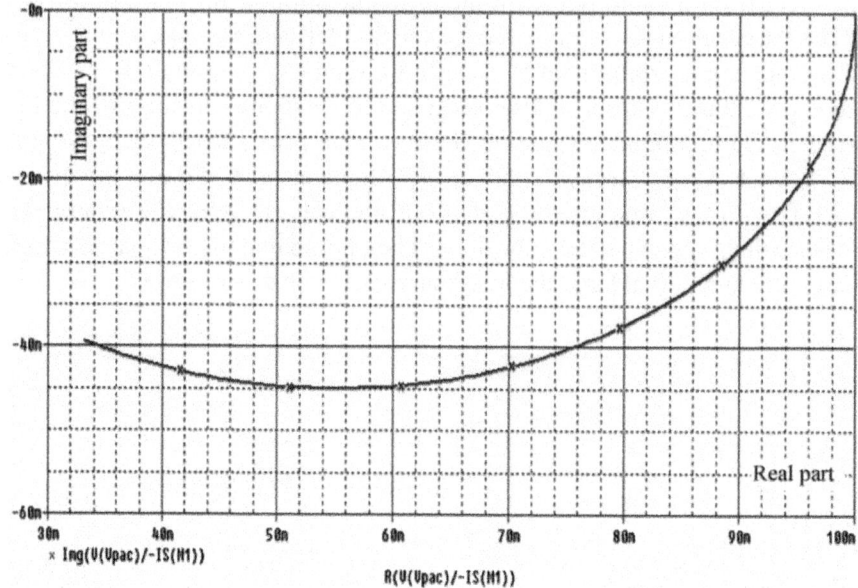

Figure 2: Imaginary part (Z') of the complex impedance (Z*) versus real part (Z'), Nyquist plot simulation.

EXPERIMENTAL STUDY

Method

The Electrochemical Impedance Spectroscopy (EIS) has the advantage, compared to other methods, to have a less influence on battery or fuel cell during the working of these latter. It can provide more information on the state of the charge. Measurements are generally carried out without load. It is useful to cover a large frequency range in order to obtain more information from the impedance spectrum generated. For a PEM fuel cell, the impedance spectrum was generated in a frequency ranging from 1Hz to 10 kHz [4]. However, Walkiewicz and al [5] did studies between 1mHz and 65KHz. The number of points collected by decade varies between 8 and 10 points. The principle of measurement is to add a signal, at constant frequency, to the output of the voltage of the battery when this latter is delivering the desired current. The superimposed signal can be obtained by three methods: potentiostatic, galvanostatic or load modulation methods.

Among of these three later methods, we have selected the load modulation method. It consists in varying the resistance of the load according to the signal that we would like to superimpose. Thus, the impedance of battery or fuel cell under test can be obtained by the ratio of the voltage of the battery and the current coming from the battery. Figure 3 shows an electric representation of this method.

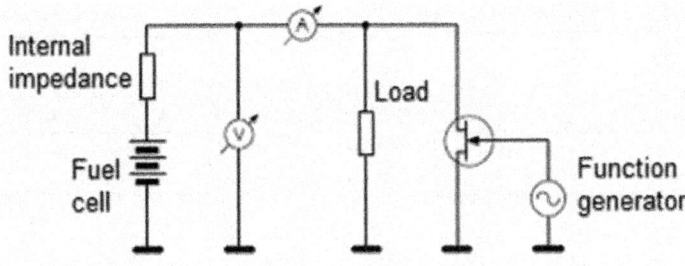

Figure 3: Principle of the load modulation method [6].

Principle of the Test Bench

The principle of the measurement using the test bench developed in our laboratory is presented in Figure 4. The current is controlled by an analogical current regulation. This allows us to have a more linear, fast and reliable regulation.

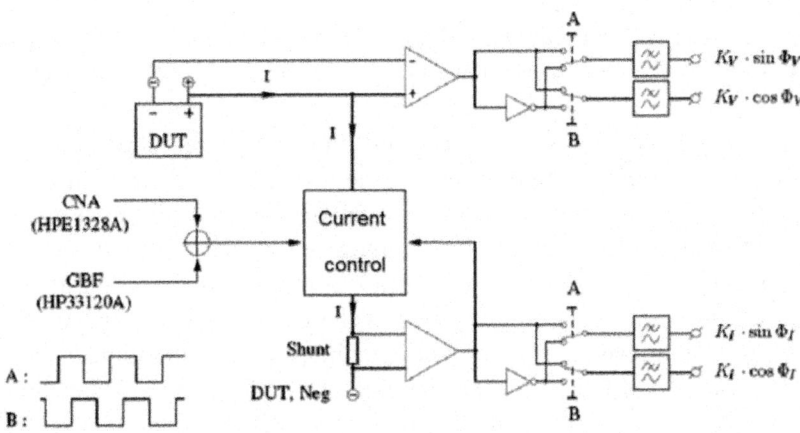

Figure 4: Synoptic of the test bench.

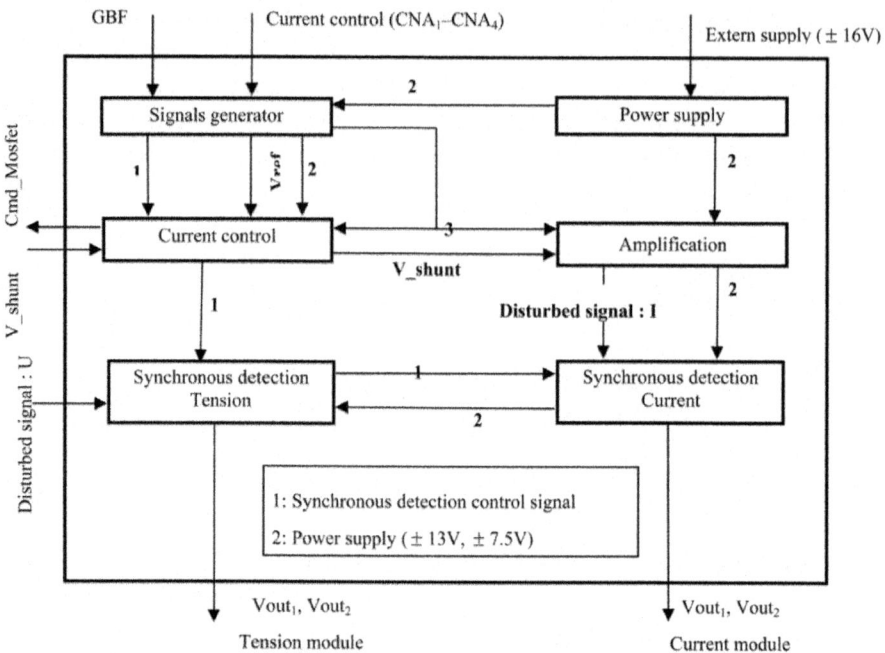

Figure 5: Systemic representation of the test bench.

The instrumentation is developed around a VXI system stand, which controls different electronic cards. Software, under Hpvee®, was developed for automatic impedance measurements of the device under test (DUT). In order to filter the noise and to avoid the influence of the 50Hz, a synchronous detection was used, which filters a very narrow way around the useful signal. Thus, it is possible to filter all the noise and to detect the amplitude of the useful signal at frequency fixed. Two synchronous detections were used: the first is used for the imposed current to the device under test, while the second is used for the response of the voltage of this same device. These two synchronous detections are controlled by four square signals delivered by an electronic card, which are out phase of 90°. The output of this synchronous detection allowed us to collect the real and imaginary part of the current and voltage, as well as, their respective phases. The real and imaginary part of the impedance of the DUT is calculated then by using the ohm›s law. These two parameters (real and imaginary part of the complex impedance) can be plotted in the Nyquist diagram.

The system developed, by our own, can support an active current up to 50A on the load. The new achievement in this work results in the possibility to better understand and to study the fuel cell in its environment when it is delivering current on load such as electric motor. In this case, the measurement and the analysis of the impedance are a good tool, which can give useful information about the state of the battery. For a better comprehension of the system operation, we make a more detailed description below.

The instrumentation is controlled by an Hpvee® program developed for this project. This program controls a VXI system stand containing several measuring devices in the form of plug-in circuits: a GBF (HPE1340A), a multimeter (HPE1326B), a 4-Channel D/A Converter (HPE1328A), a multiplexer 16 ways (HPE1351A) and an input/output circuit (HPE1330B). The "Power supply" module provides the supply to the various electronic circuits, it delivers a tension of ±13V and ±7.5V. The "Signals generator" module provides the control square signals for the synchronous detections of the tension and the current, as well as the current imposed signal. These signals are generated from the sinusoidal signal delivered by the GBF. The frequencies scanning is controlled by the Hpvee® program, which changes the frequency value step by step defines after the measurement of $Vout_1$ and $Vout_2$ of the tension and the current. The "Current control" module drives the load by an imposed current while running. This current is a square signal, generated by the "Signals generator" module; it is composed of a DC part which represent the imposed current and an AC part which represents the frequency on which this current is imposed. The "Amplification" module amplifies the imposed current signal, measured at the Shunt resistance terminals. This amplified signal which is disturbed is transmitted to the synchronous detection (current). The "Synchronous detection" module allows the amplification of the signal coming from the DUT, and the recovery of the real and imaginary part of this signal by the synchronous detection.

Figure 6: Test bed for the impedance measurement of a fuel cell on load.

"Power Supply" Module

The "Power supply" module in figure 7, manages the generation and the distribution of the various tensions which are necessary to supply the active elements of all the other modules. It transforms the tensions of ±16 V to a ±13 V and ±7.5 V tensions. Two diodes at the input allow the protection of the module in the event of a surge or a bad polarization.

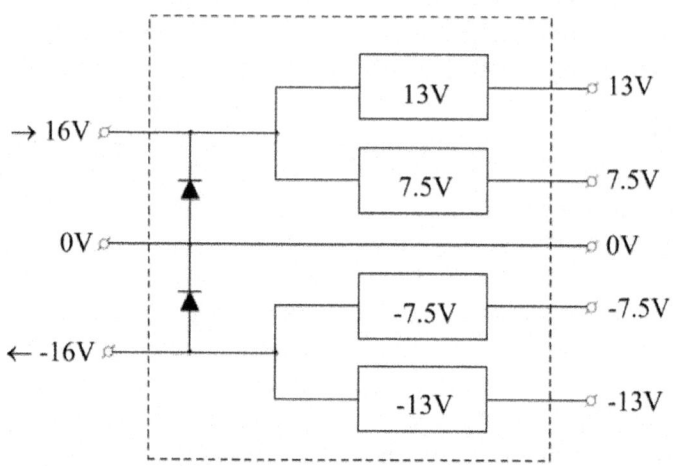

Figure 7: Schematic of the "Power supply" module.

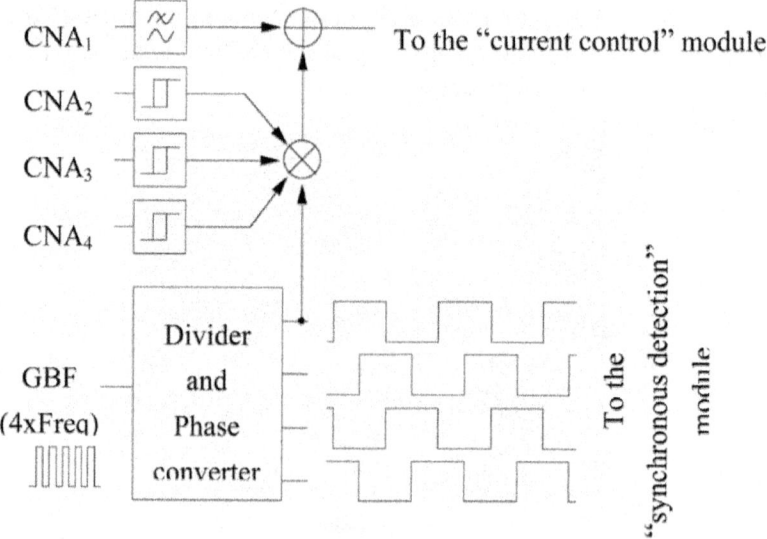

Figure 8: Schematic of the "Signals generator" module.

We decided to create this module to allow the circuit to be autonomous and to be embedded in a vehicle. Its finality being to be an embedded system, we were to avoid being dependent on any external power supply. The external supply which provides the ±16 V can be replaced by batteries. For our tests and for the measurements, we used a simple laboratory power supply. In the future, this supply must be provided by the fuel cell.

"Signals Generator" Module

The "Signals generator" module is useful like interfaces for all the input signals. It converts the square signal coming from the GBF and the tensions provided by the D/A Converter which adjusts the current and which chooses the amplification of the signal. This module amplifies and filters all the input signals. Starting from the signal provides by the GBF, it generates four square signals of the same frequency, out of phase of 0, $\pi/2$, π and $3\pi/2$. These signals allow the commutation of the quad bilateral switch (CD4066) on the synchronous detection module. The fuel cell output current will be controlled by the CNA_1 tension superimposed with one of the four outputs of the phase-converter. For that, the "Signals generator" module carries out the addition between these two signals.

The CNA_2 to CNA_4 tensions are used for the choice of the measuring range. The signal AC is not provided any more by the GBF, but by the output of the CMOS quad bilateral switches. Behind the phase-converter and before the adder, tension dividers (which we can select with three relays) determine three levels of amplification, allowing a measurement over three decades of impedance.

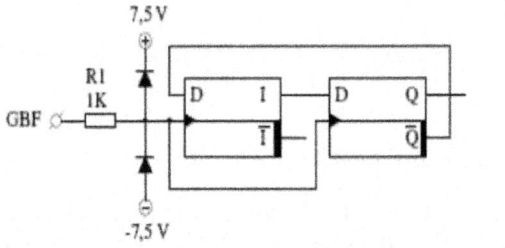
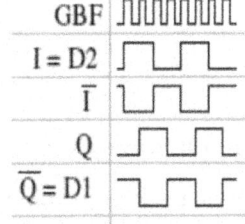

a) The phase-converter

b) Output signals driving the synchronous detection module

Figure 9: The phase converter and its outputs signals.

"Current Control" Module

This module regulates the current output by the fuel cell. It is a circuit which appears on page 382 of the book "The Art of Electronics" [10]. The operational amplifier regulates the current going from DUT,POS to DUT,NEG via the MOSFET according to the V_{ref} tension. The adjustments and the choice of the components are made to have for a V_{ref} tension of 1V, an equivalent current of 1A. The tension measured on the shunt resistance terminals, is the image of the imposed current. 100mV of this tension corresponds to a current of 1A. The tension measured at the terminals of the DUT corresponds to the response in tension of the DUT to the imposed current. These two tensions will enable us to define the impedance of the DUT by using the Ohm›s law, and that by making the extraction of the real and imaginary part of the current and the tension.

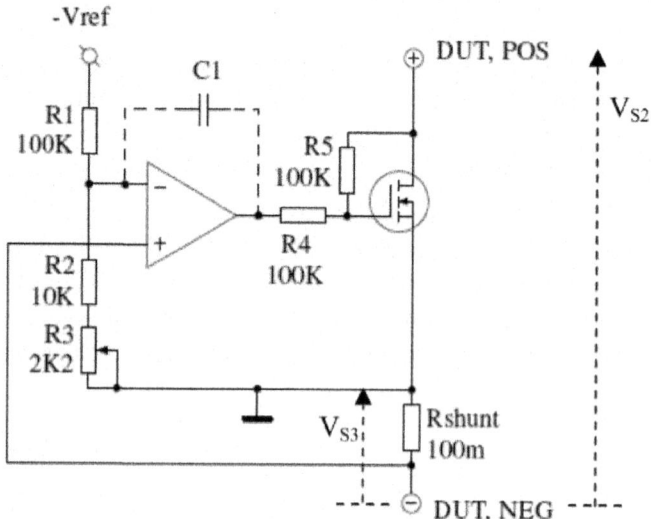

Figure 10: Schematic of the "Current control" module.

We can observe below in figure 11, the results obtained by the current control module. Figure 11.a and figure 13.b show the signal (V_{S3}) obtained at the shunt resistance terminals. We have a square signal which corresponds to the current image with an amplitude of 400mV corresponding to a 4A current. Figure 11.b and figure 13.b show the signal (V_{S2}) obtained at the DUT terminals which corresponds to the response in tension of the imposed current. It has an amplitude of 200mV. With these two parameters, we can determine the real part of the impedance by dividing the tension by the current. In this case and with this measurement values, the real part is of 50mΩ.

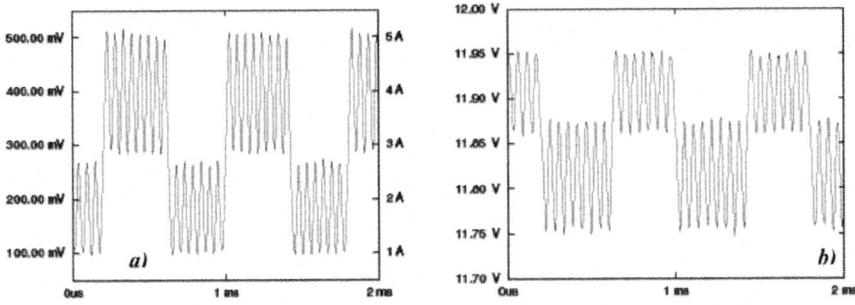

Figure 11: a) Image of the current imposed to the DUT (100mV = 1A) | V_{S3} of figure 13.b b) Response in tension coming from the DUT | V_{S2} of figure 13.b.

"Amplification" Module

This module amplifies the alternate component of the tension around the resistance of shunt before providing the signal amplified to synchronous detection. The principal constraint of this module is a weak offset of tension because this last is not filtered by synchronous detection. Moreover, amplification must be eligible in order to use all the dynamics of synchronous detection to reach a better resolution and to decrease the errors introduced by the offsets of tension. The circuit must also have an output with the reversed signal and a second with the not-reversed signal. These two signals are used for synchronous detection. The circuit is presented below in figure 12.

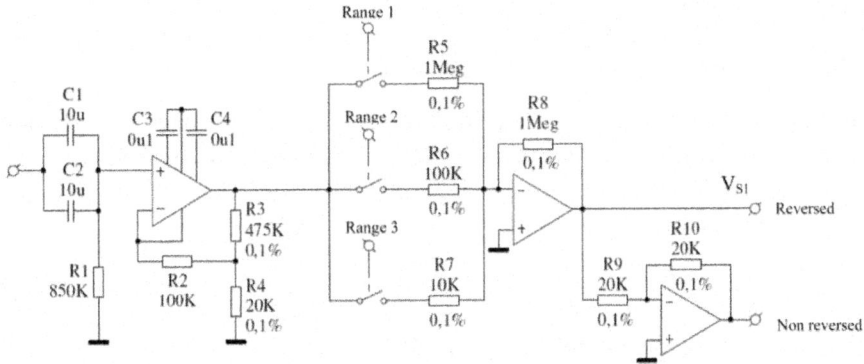

Figure 12: Schematic of the "Amplification" module.

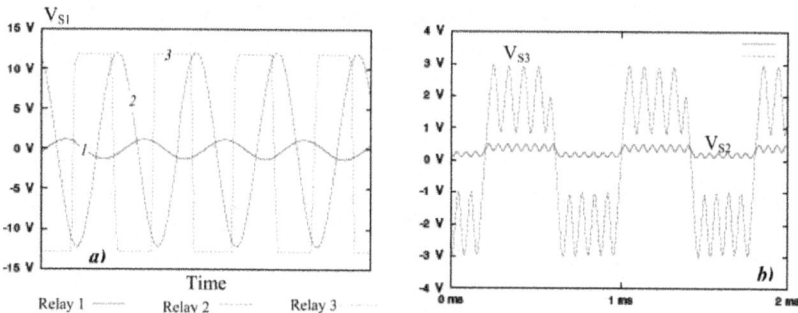

Figure 13: a) Amplification according to the position of the relay b) Amplification of the tension measured at the shunt resistance terminals

Figure 13 shows the results of measurement carried out on this module. For measurement presented on figure 13.a, the input was excited by a sinusoidal signal with an amplitude of 50mV, provided by the GBF. The amplification rate is a function of the used relay: 23 for relay 1 (range 1), 230 for relay 2 (range 2) and 2300 for relay 3 (range 3). If the third relay is open, amplification is so high that the exit becomes saturated. For measurement presented in figure 13.b, an amplification of the signal at the shunt resistance terminals is made. Its continuous component is filtered by the input capacitive of the module before an amplification by 23. It is noticed that the square signal is rounded a little compared to former measurement. This is due to the reduced band-width of the amplification module due to the Unity Gain Bandwidth of the AOP ICL7650SCPD.

"Synchronous detection" module

The synchronous detection of our test bed is composed of two basic elements. These two elements obtain the same input signal, but two different signals of reference out of phase from exactly 90°. In this case, when the first phase detector has an output signal of:

$$Vout_1 = \frac{2E}{\pi}\cos(\phi) \qquad (1)$$

the second detector presents a signal of :

$$Vout_2 = \frac{2E}{\pi}\sin(\phi) \qquad (2)$$

at its output. Considering that:

$$\sin^2(\Phi) + \cos^2(\Phi) = 1 \qquad (3)$$

we can calculate:

$$V^2 out_1 + V^2 out_2 = \left(\frac{2E}{\pi}\right)^2 \left(\sin^2(\phi) + \cos^2(\phi)\right)$$

$$\tag{4}$$

We can deduce that:

$$E = \frac{\pi}{2}\sqrt{V^2 out_1 + V^2 out_2}$$

$$\tag{5}$$

By using another law of trigonometry, we can obtain the phase of the signal:

$$\phi = \arctan\left(\frac{Vout_2}{Vout_1}\right)$$

$$\tag{6}$$

To calculate the real and the imaginary part of the signal, it is enough to know that:

$$Re(E) = E\cos(\phi) \tag{7}$$

$$Im(E) = E\sin(\phi) \tag{8}$$

Since a synchronous detection can measure only one magnitude at the same time (either the current, or the tension), we put two of them for the measurement of the current and the tension simultaneously to determine the impedance of the DUT on load.

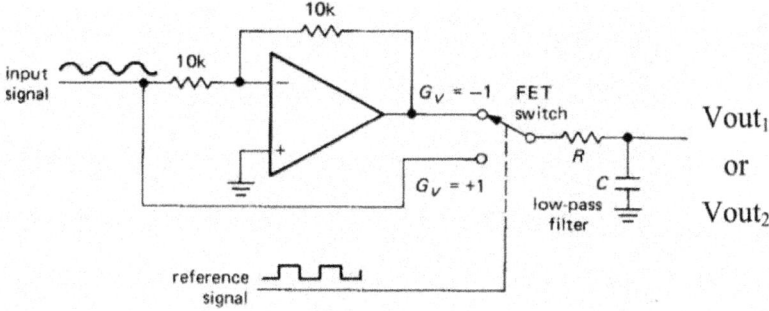

Figure 14: Principle of the synchronous detection module [10].

RESULTS AND DISCUSSION

The preliminary results have been carried out on a vehicle battery delivering a starting current of 330A and having a capacity of 40Ah (impedance measurements on a PEM fuel cell on load are in progress). The impedance

Nyquist graph of a fuel cell and a vehicle battery are very close [11][12] because the electrochemical processes are almost identical [13]. This gives us the possibility to make measurement test on a vehicle battery. However, spectrums are very depending to the values of components used in the modelization of the Randles circuit. Measurements were carried out at different nominal voltages (12.7V, 10V, 8V and 5V) with two imposed currents (0.6 A and 4A). The choice of these limits current is arbitrary. Figure 15 and 16 show the complex plane impedance plots (Nyquist diagram). The results obtained enable us to show the influence of the load on its impedance. Nyquist Graphs showed below were obtained by using the software Hpvee® developed for the opportunity it is then transposed under Microsoft® Excel in order to plot the curves. Nyquist graphs are generally presented in the literature have a positive imaginary axis. Actually, values on the axis of the imaginary part are negative (effect capacitor), but by convention, at the time of the tracing of the graph, they are multiplied by -1. In order to visualize the different phenomena and the effects occurring inside the battery, basically capacitor effect, we prefer to keep the negative imaginary axis.

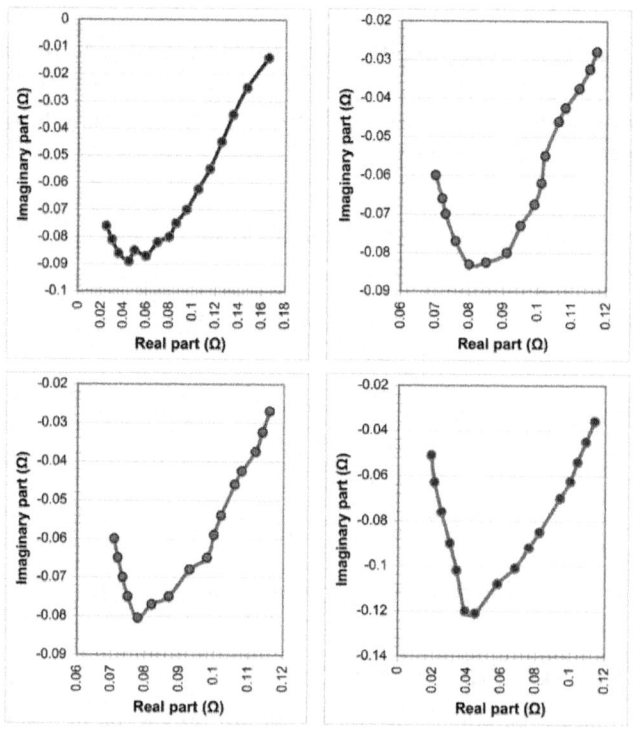

Figure 15: Impedance of the battery at an imposed current of 0.6A and at nominal voltage of: a) 12.7V b) 10V c) 8V d) 5V.

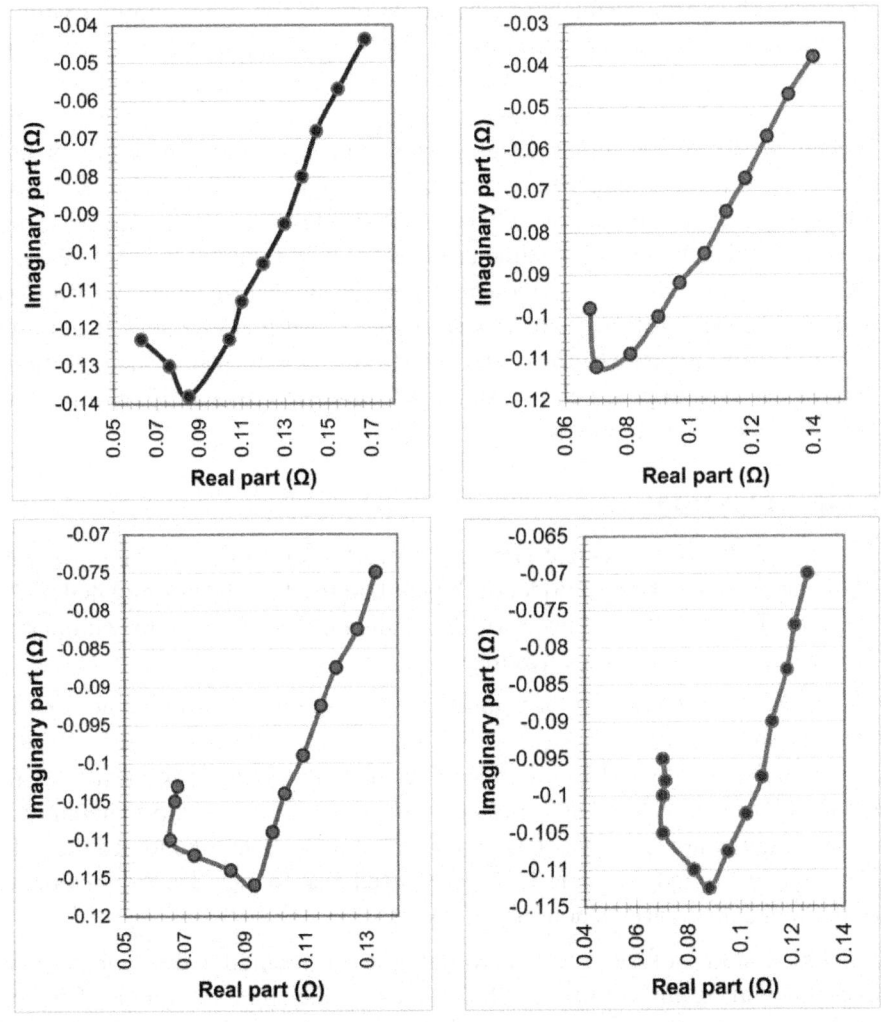

Figure 16: Impedance of the battery at an imposed current of 4A and at nominal voltage of: a) 12.7V b) 10V c) 8V d) 5V.

As it can be seen, the shape of the curves shown in Figure 15 demonstrates the ability of our system to measure the impedance of a DUT on load. The shape of the curve obtained at a nominal voltage of 12.7V and at current imposed of 0,6A is similar to the shape of the theoretical curve shown in figure 2 using the simulation. As it can be seen, the curve become more linear when the nominal voltage of the battery decreases, which means a discharge of this latter. This phenomenon can be seen for a nominal voltage of 5V (Figure 15.d). A pseudo semi-circle obtained if we do not take into account of the right stiffness. The shape of this curve could be due to the weak nominal voltage at which this measure has been made. Below a nominal voltage of 4V, our system of measure is not more capable to make some correct and exploitable measurement. This could be due to the level of tension drain/source of the Mosfets that must be important enough for measurement. The experimental curves show the predicted behavior by the theory at low frequencies. Resistive effect is generated by a positive value at the level of the real axis, while capacitor effect by negative value at the level of the imaginary axis. We can also observe a variation of component values, basically resistances of diffusion, with the discharge of the battery. The second set of measurement at imposed current of 4A (Figure 16) show that the curves have the same shape to those obtained with at imposed current of 0.6A (Figure 15), however, values of real and imaginary axes are different.

CONCLUSION

The impedance measurement is a very powerful tool for controlling the state of battery. The theoretical model of the method used has been simulated under Pspice®. This study was necessary to validate our concept by comparing theoretical and experimental results.

In the theoretical part, we gave the principle of measurement and the description of our test bench, as well as, the different electronic cards. Results of simulation reinforce us in the idea that the way that has been followed for the development of this band measurement was good. The experimental part shows also the ability of the developed system to measure the impedance of a vehicle battery, and therefore it could be used also to measure the impedance of the fuel cell at various nominal voltages.

The first aim of these tests is to validate our method and to compare the experimental results with those obtained using the simulation under Pspice®. On the other hand these testes can also confirm the choice of the method of load modulation, and the good electronic card working developed for this end. The different Nyquist graphs show that a relationship could be exist therefore between the state of load and the internal impedance of the DUT. In the case of the lead battery, as the one used in this study, the variation of the impedance is generally weak (in the order of millis ohms) in the frequency range used. The correlation between the theoretical and experimental curves can confirm that our test bench allows to measure and to plot the impedance of a battery or fuel cell in frequencies. In the future we are interesting by adding humidity sensors to be able to compare and to correlate the impedance of the fuel cell on load with the humidity level inside it. This correlation will give us informations on the membrane degradation. This equipment could be integrated in a vehicle functioning with a fuel cell in order to control the deterioration of its membranes by using data from control sensors and measurement equipments.

ACKNOWLEDGMENTS

The authors would like to thank the Lorraine region – France – for supporting this work.

REFERENCES AND NOTES

1.   Easton, E.B.; Pickup, P.G. An electrochemical impedance spectroscopy study of fuel cell electrodes. *Electrochimica Acta*.2005, *50*, 2469–2474.

2.   Jasinski, P.; Suzuki, T.; Dogan, F.; Anderson, H.U. Impedance spectroscopy of single chamber SOFC. *Solid State Ionics*.2004, *175*, 35–38.

3.   Li, G.; Pickup, P.G. Measurement of single electrode potentials and impedances in hydrogen and direct methanol PEM fuel cells. *Electrochimica Acta*. 2004, *49*, 4119–4126.

4.   *Making Fuel Cell AC Impedance Measurements Utilizing Agilent N3300A Series Electronic Loads*; Product Note for Agilent Technologies, Inc.: Santa Clara, CA, 2002.

5.   Walkiewicz, S. *Étude par spectroscopie d›impédance électrochimique de piles à combustibles à membrane échangeuse de protons. (in French)*; DEA Électrochimie, Institut National Polytechnique de Grenoble ENSEEG: Grenoble, France, 2001.

6.   Kraemer, B. *Mesure par spectroscopie de l'impédance d'une pile à combustible en charge (in French)*; DEA Rapport; UHP Nancy1: Nancy, France, 2005.

7.   Noponen, M. Current Distribution measurements and Modelling of Mass Transfer in Polymer Electrolyte Fuel Cells. Ph.D. Thesis, Helsinki University of Technology, March 2004.

8.   Wagner, N. Characterization of membrane electrode assemblies in polymer electrolyte fuel cells using AC impedance spectroscopy. *Journal of Applied Electrochemistry* 2002, *32*, 589–863.

9.   Brunetto, C.; Tina, G.; Squadrito, G.; Moschetto, A. PEMFC Diagnostics and Modelling by Electrochemical Impedance Spectroscopy. *IEEE MELECON* 2004, *3*, 1045–1050.

10. Horowitz, P.; Hill, W. *The Art of Electronics*, 2nd Ed. ed; Cambridge University Press: Cambridge, 1989.

11. Diard, J.P.; Le Gorrec, B.; Montella, C.; Poinsignon, C.; Vitter, G. Impedance Measurements of Polymer Electrolyte Membrane Fuel Cells Running on Constant Load. *Journal of Power Sources* 1998, *74*, 244–245.

12. Diard, J.P.; Le Gorrec, B.; Montella, C. EIS Study of Electrochemical Battery discharge on constant load. *Journal of Power Sources* 1998, *70*, 78–84.

13. Jörn, A.T. Multiple Model Impedance Spectroscopy Techniques for testing Electrochemical Systems. *Journal of Power Sources* 2004, *136*, 246–249.

Chapter 2

RICE: A RELIABLE AND EFFICIENT REMOTE IN-STRUMENTATION COLLABORATION ENVIRONMENT

Prasad Calyam, Abdul Kalash, Ramya Gopalan, Sowmya Gopalan, and Ashok Krishnamurthy

Cyberinfrastructure and Software Development Group, Ohio Supercomputer Center, 1224 Kinnear Road, Columbus, OH 43212, USA

ABSTRACT

Remote access of scientific instruments over the Internet (i.e., remote instrumentation) demand high-resolution (2D and 3D) video image transfers with simultaneous real-time mouse and keyboard controls. Consequently, user quality of experience (QoE) is highly sensitive to network bottlenecks. Further, improper user control while reacting to impaired video caused due to network bottlenecks could result in physical damages to the expensive instrument equipment. Hence, it is vital to understand the interplay between (a) user keyboard/mouse actions toward the instrument, and (b) corresponding network reactions for transfer of instrument video images toward the user. In this paper, we first present an analytical model for characterizing user and network interplay during remote instrumentation sessions in terms of demand and supply interplay principles of traditional economics. Next, we describe the trends of the model parameters using subjective and objective measurements obtained from QoE experiments. Thereafter, we describe our Remote Instrumentation Collaboration Environment (RICE) software that leverages our experiences from the user and network interplay studies, and has functionalities that facilitate reliable and efficient remote instrumentation such as (a) network health awareness to detect network bottleneck periods, and (b) collaboration tools for multiple participants to interact during research and training sessions.

INTRODUCTION

Increased access to high-speed networks has made remote access of computer-controlled scientific instruments such as microscopes, spectrometers, and telescopes widely-feasible over the Internet. Some of these instruments are extremely expensive and could be worth several hundred-thousand dollars. Hence, a major benefit of remote instrumentation is that it allows remote users to utilize these instruments when they are not in use by local users. In addition, routine maintenance and operation of the instruments require significant investment in staffing. Thus, instrument labs can charge remote access on an hourly usage basis to obtain a better return-on-investment on the instruments. Further, remote instrumentation avoids duplication of investment in instrument labs for funding agencies. In fact, the National Science Foundation is mandating remote instrumentation to be available with all their funded instruments [1]. Besides the above advantages, remote instrumentation fosters education and hands-on training of instruments as well as collaboration for remote users. The collaboration enables multiple remote researchers, each with unique expertise, to jointly analyze samples such as metals, proteins, and tissues. All of the above advantages, especially for training and collaboration, drastically shorten the development process involved in innovations related to materials modeling, biological specimens' analysis for cancer research, and so forth. At the same time, they improve user convenience and significantly reduce research and training costs.

Although there are several advantages, remote instrumentation is demanding in terms of network resource consumption. This is because remote instrumentation sessions involve high-resolution (2D and 3D) video image transfers with simultaneous real-time mouse and keyboard controls. If appropriate network bandwidth is not allocated, network congestion occurs that can impact user quality of experience (QoE). In addition, user QoE is affected by network fault events such as optical fiber cuts, route asymmetry, and route flapping that degrade network performance. The user QoE affected by such network bottlenecks is measured by obtaining subjective opinions of user satisfaction after completion of a remote instrumentation session. The network bottleneck could cause impaired video images at the user, which in turn could lead to improper user control of the microscope's mechanical moving parts. Such improper user control may ultimately result in physical equipment damages that are prohibitively expensive to fix. Hence, it is vital to understand the interplay between (a) the user keyboard/mouse actions toward the instrument, and (b) the corresponding network reactions for transport of instrument video images toward the user, for reliably supporting remote instrumentation.

Assuming that a sample has been shipped to an instrument lab and has been loaded into an instrument, there are two basic use-cases of remote instrumentation. The first use-case is called remote observation, where a remote user or multiple remote users only view real-time (2D and 3D) instrument video images. The remote user(s) direct an operator physically present at the instrument to perform all the control actions over a telephone or VoIP call. The second use-case is called remote operation, where a remote user or multiple remote users view the instrument video images and also control the instrument in real-time. The first use-case is preferred in cases where the remote users are not familiar with the instrument functionalities. It is also preferred if the intermediate network path between the user and the instruments has bottlenecks. The second use-case is preferred for both local and remote users in cases where human presence around the sample could cause undesirable effects. For example in microscopy involving electron microscopes, human presence increases ambient temperature, which alters properties of materials being analyzed at subangstrom levels on the microscope. Nevertheless, both use-cases require collaboration tools that support voice communications (i.e., VoIP) and instant messaging (i.e., chat) for communicating efficiently during remote instrumentation sessions. For the multiuser case, collaboration tools are required to (i) show who is controlling/viewing the session (i.e., presence) and (ii) manage control privilege amongst the users (i.e., control-lock passing) such that at any given instant, only one user controls the instrument.

There are two major parts to this paper. In the first part, we study the complex interplay characteristics between the user and the network during remote instrumentation sessions. As an exemplar for the interplay characterization, we focus on the remote access of electron microscopes (i.e., remote microscopy). However, our work is equally relevant for other computer-controlled scientific instruments. We first present an analytical model for characterizing user and network interplay during remote microscopy sessions in terms of demand and supply interplay principles of economics, respectively. The various remote microscopy system states affected by transient network conditions are also modeled. To obtain the trends of the session model parameters, we set up a remote microscopy testbed in cooperation with The Ohio State University's Center for Accelerated Maturation of Materials (CAMM). On this testbed, we use a novel methodology to perform QoE experiments involving actual novice/expert users for a variety of network conditions in

LAN/WAN connections. We also present the analysis of the subjective and objective measurements obtained from the experiments. Our analysis provides insights about how network health impacts user behavior and ultimately user QoE.

In the second part, we describe our Remote Instrumentation Collaboration Environment (RICE) software that leverages our user and network interplay study findings to (a) cope with network bottlenecks, and (b) cater to the multiuser requirements of remote observation and remote operation. In this context, we describe the RICE software functionalities that improve reliability and efficiency of multiuser remote instrumentation sessions that traditionally relied upon off-the-shelf virtual network computing (VNC) solutions [2]. The functionalities to improve reliability include real-time network health monitoring coupled with network performance anomaly detection using a "plateau-detector algorithm." This algorithm warns and blocks user's control actions during network congestion periods. We also describe our "session-signaling protocol" used in the RICE tools that improve efficiency of multiuser collaboration. The collaboration tools include VoIP, chat, presence, and control-lock passing that are absent in off-the-shelf VNC solutions. Finally, we present potential applications of RICE for research and training purposes that require multiuser remote instrumentation capabilities.

The remainder of the paper is organized as follows. Section 2 presents related work. Section 3 presents the analytical model for user and network interplay characterization. Section 4 describes the remote microscopy testbed and results of the QoE experiments performed on the testbed. Section 5 describes the RICE software features and its applications for research and training. Section 6 concludes the paper.

RELATED WORK

There are several efforts in the United States that are aimed at serving the remote instrumentation needs of researchers and students. Gemini Observatory [3] is an initiative that uses Internet2 to allow remote users to manipulate their twin telescopes. NanoManipulator [4] is another initiative that uses Internet2 to allow remote control and visualization of images from their scanning probe microscopes. Similar remote instrumentation efforts are being supported in other countries also. A notable effort is being led by the National Institute of Materials Science in Japan, where remote instrumentation is being made available to the public and high school education programs [5]. As a part of this effort, remote observation of insects, plants, IC devices, and metals that have been preloaded in a remote-site's scanning electron microscope is being enabled at the National Museum of Emerging Science and Innovation in Tokyo.

One of the early works that developed novel applications for remote operation were done at the Massachusetts Institute of Technology [6]. They developed a custom software for remote control of a Zeiss microscope using a graphical interface running on a workstation computer. The software also allowed several remote users to simultaneously view the microscope in a conference inspection mode, enabling collaboration amongst remote users. Lawrence Berkeley National Laboratory also has developed a custom software application to control their Kratos 1500 keV microscope during in situ experiments [7]. The controls include adjusting external stimuli, adjustment of specimen position and orientation, and manipulation of microscope controls such as illumination, magnification, and focus. To cope with network bottlenecks, they developed schemes to locally automate stage control and microscope focus. Their application has been tested on the Internet along several paths including paths to Berkeley from Washington D.C. and Kansas City. Several other studies have also evaluated performance of remote instrumentation using custom software over the Internet. For example, Research Center for Ultra High Voltage Electron Microscopy (UHVEM) at Osaka University collaborated with National Center for Microscopy and Imaging Research (NCMIR) at University of California San Diego to conduct remote instrumentation experiments on their 3-million volt transmission electron microscope over intercontinental links [8]. The custom software developed by NCMIR has evolved over the years to keep up with the developments of networks, operating systems, and application development tools. The latest variants of their software feature platform-independent Java-based applications for remote instrumentation of several different instruments. These applications have also been integrated into web-services and middleware frameworks [9] that couple remote instrumentation with data and computation services.

Recently, several off-the-shelf remote access solutions have emerged that are either software-based or hardware-based. The most commonly used solution is the software-based virtual network computing (VNC) solution [2] that has several variants such as UltraVNC [10] and RealVNC [11]. It requires preinstalled software at both the instrument and user ends. Alternately, there are hardware-based VNC solutions that are also referred to as Keyboard, Video and Mouse over IP (KVMoIP) solutions developed by vendors such as ThinkLogical [12] and Avocent [13]. These solutions use custom hardware and require a pair of encoder and decoder appliances to be installed at the instrument and user ends. Recently, hybrid VNC solutions have also been developed by vendors such as Adder [14] that requires a hardware appliance at the instrument end, and a software client at the user end. Several instrument labs use such off-the-shelf solutions. For example, Oak Ridge National Laboratory uses off-the-shelf VNC solutions for remote control of their High Flux isotope Reactor [15]. Similarly, the California State Polytechnic University also uses off-the-shelf VNC solutions in their Ocean Engineering Program [16].

VNC solutions use raw or copy-rectangle or JPEG/MPEG encoding for video image transfers and TCP for keyboard and mouse control traffic. For sending the video image transfers, VNC uses a Remote Frame Buffer (RFB) protocol that supports various pixel formats such as ZRLE, Zlib, Raw, and Hextile. The pixel updates using the RFB protocol are demand-driven because pixel updates are sent (a) to respond to an explicit TCP-based request from a client, and (b) to update the client's display when there are changes at the server's display. The compression latency of VNC is dependent on factors such as the network health, as well as the client/server CPU speed, other-application task loads, and video card capabilities.

Given the free availability of software-based VNC solutions, QoE evaluations for these solutions can be extensively found in the remote instrumentation literature. However, to the best of our knowledge, there is no literature on systematic QoE evaluations for KVMoIP-based remote instrumentation. The QoE evaluation results presented in Section 4 of this paper focus on the KVMoIP VNC solution on both LAN and WAN paths. We also believe that our work is the first to present an analytical model and characterize user and network interplay in remote instrumentation sessions. Our RICE software presented in Section 5 is based on the UltraVNC solution but has several enhancements targeted for network-aware and collaborative remote instrumentation sessions. Our work in this paper is part of The Ohio State University's CAMM VIM program [17]. This program uses Ohio Supercomputer Center's (OSC) regional network (i.e., OSCnet) to allow remote industry such as Timken and defense labs such as AFRL to access their collection of the world's most powerful scanning/transmission electron microscopes.

REMOTE MICROSCOPY SESSION MODEL

In this section, we first describe the parameters involved in a typical remote microscopy session. Next, we model their interactions in different system states borrowing the supply and demand terminology from economics.

System Description

Figures 1(a) and 1(b) show a basic remote microscopy system and its closed-loop control system representation with the different session parameters, respectively. The remote user physically controls the functions of the microscope by interacting with a graphical user interface (GUI) application using keystrokes and mouse moves/clicks via VNC or KVMoIP (console). Examples of microscope functions include adjusting stage position, lens focus, and magnification levels. The GUI application actually resides on a computer directly connected to the microscope's video output and control input ports. Figures 2(a) and 2(b) show two distinct video activity levels (i.e., temporal and spatial characteristics in the GUI application images sent from the microscope to the remote user). We can notice that the images contain live video feeds of instrument cameras with high video activity levels, or text and graphs with low video activity levels.

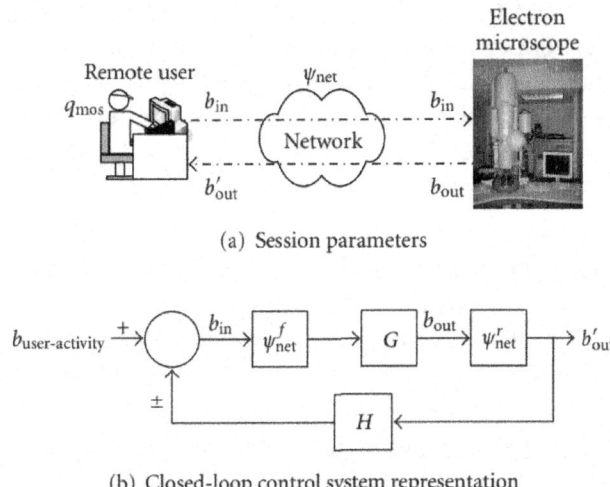

(a) Session parameters

(b) Closed-loop control system representation

Figure 1: Basic remote microscopy system.

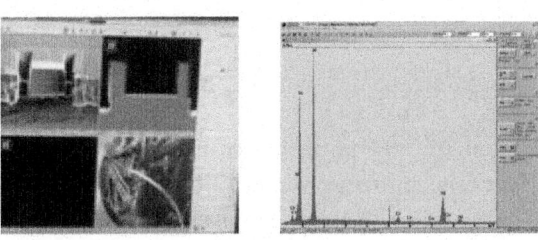

Figure 2: Comparison of video activity levels in instrument image transfers.

Let b_{action} be the average bit rate of the TCP control traffic that is generated due to keystrokes and mouse moves/clicks at the user end to accomplish a particular microscope function. The user-activity input to the system during a session involving n microscope functions can be denoted by $b_{\text{user-activity}}$ given in

$$b_{\text{user-activity}} = \sum_{i=1}^{n} \left(b_{\text{ith action}}\right).$$

(1)

For such an input, the average video image transfer rate (i.e., RTP media traffic output at the microscope end (b_{out})) can be denoted as follows:

$$b_{\text{out}} = \psi_{\text{net}}^{f} G(b_{\text{in}} + b_{\text{seed}}),$$

(2)

where ψ_{net}^{f} is the network connection quality between the user and the microscope. The network connection quality refers to the end-to-end throughput that is affected by network congestion and network fault events. The G corresponds to the input-output scaling factor which is unique for a microscope function. The b_{seed} corresponds to the rate at which periodic intracoded frames (I-frames) are sent from the encoder (at the microscope) to the decoder (at the user) for quick image refresh upon recovery from network partition events during a session.

Although b_{out} is sent from the microscope, there are two network factors that could degrade the average video image transfer rate at the user end (b_{out}). The first factor is the network connection quality of the reverse path (i.e., between the microscope and the user (ψ_{net}^{r}). The second factor is the available bandwidth in the intermediate network path. As shown in the following equation, if adequate available bandwidth is provisioned, b_{out}' will be equal to b_{out}; otherwise, b_{out}' is limited to b_{net}, which refers to the bottleneck hop bandwidth:

$$b_{\text{net}} = \min_{i=1..\text{hops}} b_{\text{ith hop}},$$

$$b_{\text{out}}' = \min (b_{\text{snd}}, b_{\text{net}}).$$

(3)

The degradation of b_{out}' manifests to users as video signal impairments such as frame freezing, blurriness, and tiling [18]. Based on the positive or negative b_{out}' feedback received at the user end from the microscope, the subsequent user behavior determines the session state. We refer to this system-state control parameter that is dependent on the user behavior as H. Details of how H parameter impacts the different system states are described in Section 3.2. We can thus express b_{in} as follows:

$$b_{in} = b_{\text{user-activity}} - H b'_{out}.$$

(4)

Using substitutions in (1)–(4), we can derive the closedloop transfer function in the classical form as shown in the following equation; this function fully describes the order, type, and frequency response for a remote microscopy system:

$$\frac{b'_{out}}{b_{\text{user-activity}}} = \frac{G\psi_{net}^{f}\psi_{net}^{r}}{1 \pm G\psi_{net}^{f}\psi_{net}^{r}H}.$$

(5)

Ultimately at the end of a session, the overall user QoE (q_{mos}) will depend on both the effort a user had to expend to perform n actions (i.e., $b_{\text{user-activity}}$) and the perceivable video image quality (i.e., b'_{out} during those actions). Hence, q_{mos} can be expressed as follows:

$$q_{mos} = f\left(\underbrace{b_{\text{user-activity}}}_{\text{Demand}}, \underbrace{b'_{out}}_{\text{Supply}} \right).$$

(6)

From (6), we can make an analogous comparison of $b_{\text{user-activity}}$ and b'_{out} to the "demand" and "supply" terminology used in economics, respectively. In traditional economics, an increase in demand levels for a commodity causes an increase in supply levels of the commodity. This in turn increases the demand, as the increased supply in large numbers generally drives down the overall commodity price. As long as both the demand and supply increase hand-in-hand by deriving reinforcement from each other, the economy (analogous to q_{mos}) is considered to be in a productive state. However, this is not always the case in the demand and supply reinforcement effect seen in remote microscopy with respect to q_{mos}. The overall network health in both the forward and reverse paths (ψ_{net}) adds complexity in the relationship of the demand and supply variables as elaborated in the next subsection, which severely affects the q_{mos}.

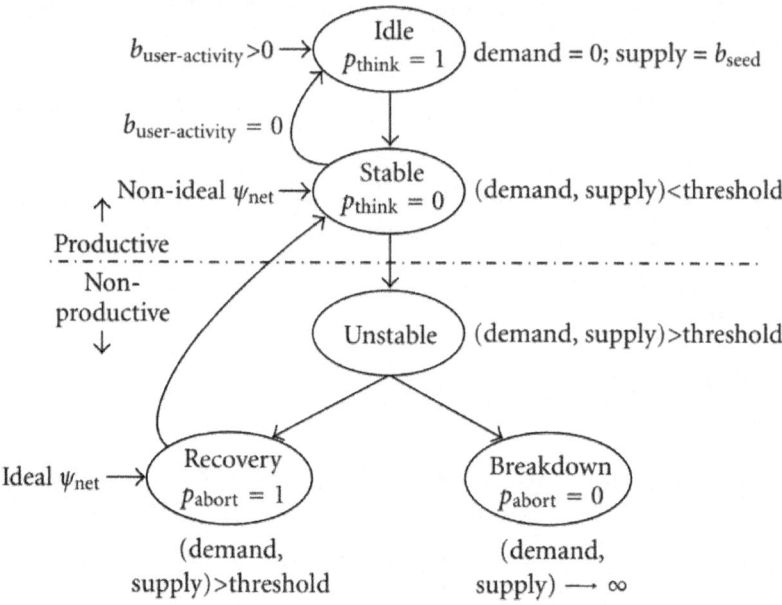

Figure 3: Remote microscopy system state transitions.

As a note, the above remote microscopy session model can be applied for both remote observation as well as remote operation use-cases. Recall that the remote operation use-case employs inband TCP control traffic toward the microscope, whereas remote observation use-case employs an out-of-band voice channel (e.g., a telephone) for directing control messages to a local user at the microscope. If we assume that a reliable voice channel exists between the two users and that the local user is responsive enough that the remote user does not perceive annoying control delays, the model is identical for both the use-cases.

System States

We now explain the interactions of the remote microscopy session parameters due to user behavior that affect the H parameter. The changes in the H parameter influence the ± sign (positive or negative feedback) of the denominator in (5) which in turn causes the different system state transitions shown in Figure 3. Initially, the system is in the "Idle" state when the user is inactive with a probability p_{idle} and the microscope GUI application is operational. In the Idle state, the demand is zero and the supply equals b_{seed} as shown in Figure 4. The remote microscopy session begins upon user-activity, and the demand and supply steadily increase. Assuming ideal ψ_{net} conditions at a given time t, the system attains a "Stable" state where the demand and supply are below the system's optimum performance threshold point (s_0, d_0). In this state, the user is successfully controlling the microscope functions and is being productive. We can now say that H is causing negative feedback in the system. At random times in this state, it is possible that a user will still be in session but idle in terms of control, presumably due to a thought process driven by a visual inspection of a sample's area of interest. Such an inactive user behavior brings the system back to its Idle state where the system is still productive. During such user inactivity times under ideal ψ_{net} conditions, we refer to p_{idle} as p_{think}

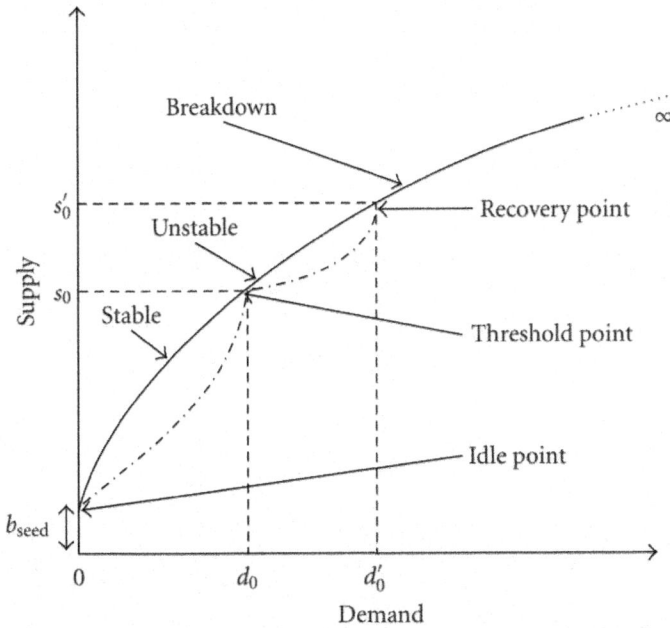

Figure 4: Remote microscopy system performance at different supply and demand conditions.

as follows:

$$p_{idle}(t) = \begin{cases} p_{think}, & \text{if ideal } \psi_{net}(t), \\ p_{abort}, & \text{if nonideal } \psi_{net}(t). \end{cases} \tag{7}$$

If the ψ_{net} were to change to nonideal conditions due to network bottlenecks caused by network congestion and network fault events, the system would enter an "Unstable" state. Here, the demand and supply rapidly increase beyond the system's optimum performance threshold point. This is because the user in this system state experiences QoE degradation effects (e.g., frame freeze) that force him to misjudge his control actions that result in unwanted supply. This is subsequently followed by a retry of the previous actions before the unwanted supply transfer completes, which further increases the demand and the QoE degradation effects and so on.

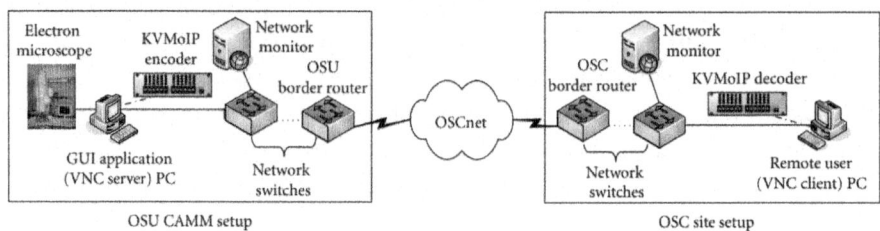

Figure 5: Remote microscopy testbed setup

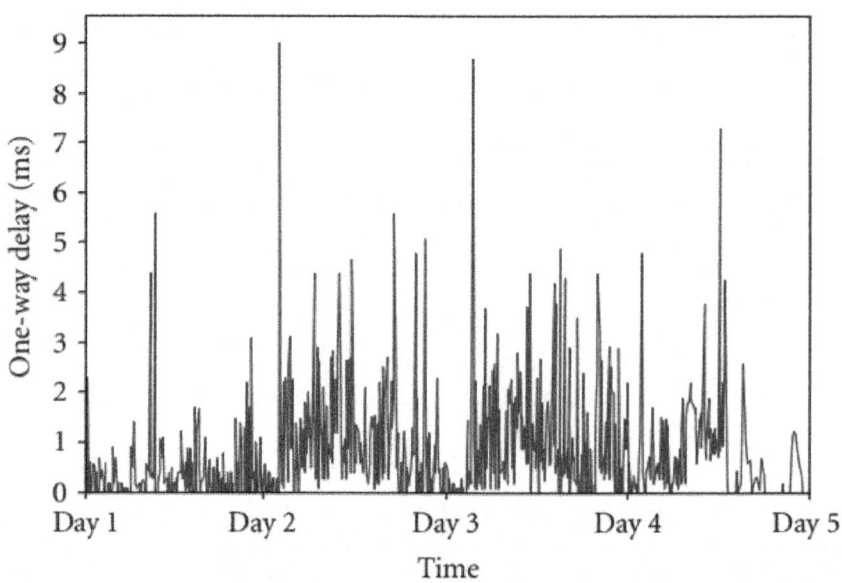

Figure 6: One-way delay between OSC and OSU CAMM.

Soon, the system becomes nonresponsive to the increasing demand, and is pressured into handling large volume of unwanted supply that is introduced from the microscope end. It is important to note that although the demand

and supply rapidly increase hand-in-hand beyond the threshold point, the system is nonproductive. We can now say that H is causing positive feedback in the system. If the user persists in his retry demand behavior, the system soon advances to a "Breakdown" state where the demand and supply tend to ∞. However, if the user aborts any actions and becomes idle at a recovery point (s'_0, d'_0), the system transitions into a "Recovery" state. During such user inactivity times under nonideal ψnet conditions, we refer to pidle as pabort as shown in (7). During the Recovery state, the demand and supply gradually tend toward the system's optimum performance threshold point. Once the ψ_{net} returns to ideal conditions (e.g., due to reduced network congestion or stabilization of the impulsive demand and unwanted supply), the system regains its Stable state and becomes productive again.

REMOTE MICROSCOPY TESTBED EXPERIMENTS

In this section, we first describe the remote microscopy testbed used to obtain trends of the different session model parameters under different network conditions. Next, we explain the test cases and performance measurements collected during the QoE experiments on the testbed. Lastly, we discuss the QoE experiment results.

Testbed Setup

For setting up the remote microscopy testbed, we collaborated with The Ohio State University's Center for Accelerated Maturation of Materials (OSU CAMM). The testbed featured four different network connections between the remote user console and the GUI application PC: (i) Direct GigE (ii) Isolated LAN (iii) Public LAN, and (iv) WAN. The Direct GigE connection had a Cisco GigE switch connecting the GUI application PC and the remote user console, which were in adjacent rooms. This connection represents the setup for avoiding users to be physically present at the microscope, especially when human presence around a sample is undesirable as explained in Section 1. The Isolated LAN connection was setup by including the CAMM's Cisco Catalyst 2924 switch to the Direct GigE connection. This connection represents remote microscopy for users in the same LAN as the microscopes, but in different lab rooms. The Public LAN connection was setup by including three additional Cisco Catalyst 2924 switches located at

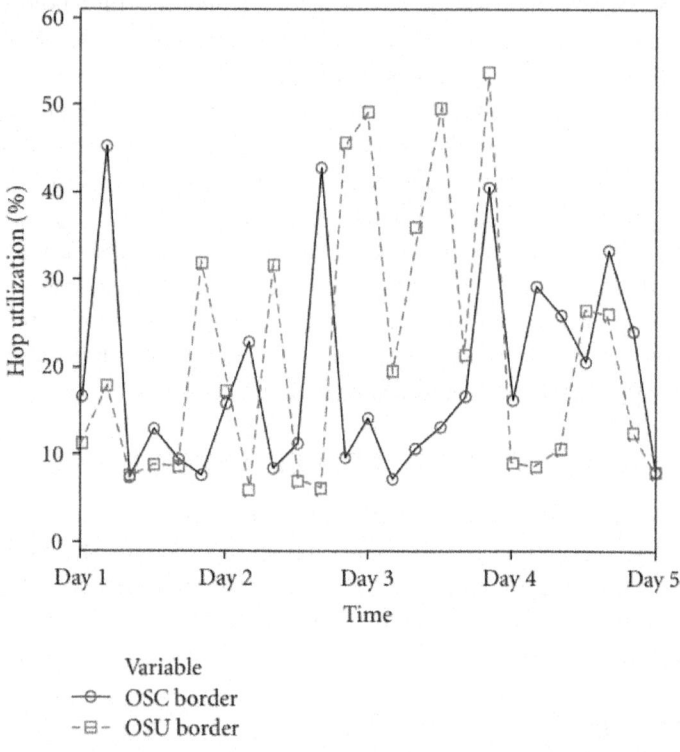

Figure 7: Border hop utilization at OSC and OSU.

neighboring buildings to the Isolated LAN connection. This connection represents remote microscopy for users working from different LANs and different lab rooms. Finally, the WAN connection was setup as shown in Figure 5 via OSCnet between OSU CAMM and OSC. This connection represents remote microscopy for users at remote sites on the Internet.

The Direct GigE, Isolated LAN, and Public LAN connections were 100 Mbps switched full-duplex connections. To know the baseline performance of the 100 Mbps WAN connection, a number of measurements were collected over a 5-day period using the OSC-developed ActiveMon software [19]. The measurements indicated the available bandwidth, delay, jitter, and loss trends. Figure 6 shows the one-way delay measurements in the path between OSC and OSU (i.e., remote user to microscope) as measured by the OWAMP tool [20] in the ActiveMon measurement toolkit. We can observe that the one-way delay measurements are generally within 10 microseconds. Figure 7 shows the bandwidth utilization levels (sampled once every 4 hours) at the bottleneck hops

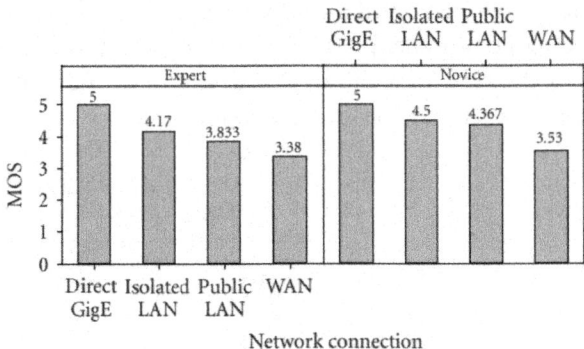

Figure 8: MOS (qmos) rankings comparison for varying ψnet conditions.

(i.e., the border routers at OSU and OSC). We can observe that worst-case utilization level is ≈50% and thus, in general, there is at least about 50 Mbps available bandwidth in the WAN path. The other routers in the WAN path are the OSCnet routers with utilization levels

Test Cases and Measurements

The test cases involved performing preassigned tasks by actual users in remote microscopy sessions using a KVMoIP VNC solution [12] over the different network connections. Raw pixel format with copy-rectangle video encoding was used in all the test cases. The actual users (i.e., human subjects) were classified under two groups: (i) "novice" and (ii) "expert," with three human subjects in each group. The novice users performed a set of sequential tasks with simplistic actions: Task-1: move view from one location on the surface of sample material to another location, Task-2: focus on high-resolution imaging, and Task-3: change the quad-screen to a single screen and grab a high resolution image. The expert users performed a set of sequential tasks with advanced actions that require relatively more effort and skill: Task- 1: eucentric height adjustment—stage movement in the Zdirection, Task-2: beam modulation—column alignment for best image, and Task-3: focus for high-resolution imaging.

During execution of the test cases, both objective and subjective measurements were collected. The objective measurements correspond to passive measurements of the control traffic (b_{in}) and video traffic (b'_{out}) collected using the popular TCPdump packet sniffing tool. The subjective measurements are the user QoE measurements, which are collected using the popular mean opinion score (MOS) ranking technique [21]. In this technique, at the end of each test case (i.e., remote instrumentation session task), the user is asked to rate his/her perceived QoE (q_{mos}) on a subjective scale of 1–5, with [1, 3) range being Poor grade, [3, 4) range being Acceptable grade, and [4, 5] range being Good grade. In addition to the MOS rankings, completion times (T) of novice and expert sessions were also recorded.

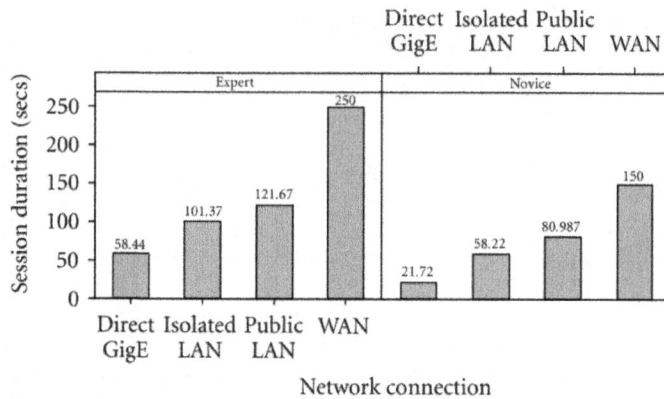

Figure 9: Session duration (T) comparison for varying ψ_{net} conditions.

QOE EXPERIMENT RESULTS

Network Connection and User QoE

First, we analyze the impact of network connection quality on the user QoE MOS (i.e., q_{mos}) in a remote microscopy session. Figure 8 shows the average qmos comparison between the novice and expert users for varying ψ_{net} conditions, with ψ_{net} being the highest for Direct GigE, and the lowest for WAN. For both types of users, we can observe that the q_{mos} rankings decrease notably with decreasing ψ_{net}. The q_{mos} equal 5 rankings of the novice and expert users while using the direct GigE connection indicates "at-the-microscope" QoE. Expectedly, for the other network connections, we can see that novice rankings are relatively more liberal than expert rankings due to the inherent intensity of the actions involved. The q_{mos} in the case of Isolated LAN and Public LAN are comparable. Further, the q_{mos} rankings for the WAN connection are in the acceptable grade, suggesting that user QoE in remote microscopy is highly sensitive to network congestion.

The average time to complete a set of predetermined user actions (i.e., session duration T) is another useful metric that provides insight about the user QoE. Figure 9 shows the session durations of the novice and expert for varying ψ_{net} conditions. For both types of users, we can observe that the session duration T increases with the decrease in ψ_{net}. For instance, the session duration T more than doubles in the Isolated and Public LAN connections in comparison to the Direct GigE connection. In the same context, we can see that the MOS ranking dip is higher for the expert (from 5 to 4.17) than the novice (from 5 to 4.5). We can generalize this observation of lesser MOS rankings for higher session durations across the other cases of both expert and novice users by comparing the session durations in Figure 9 with the MOS rankings in Figure 8. The longer session durations can be attributed to the additional effort (e.g., mouse moves/clicks, keyboard strokes, waiting for image transfer) involved while coping with network health fluctuations. Given that the difficulty level of expert user tasks is higher than novice user tasks, we can see that MOS ranking dip is higher in the expert user cases than the novice user cases.

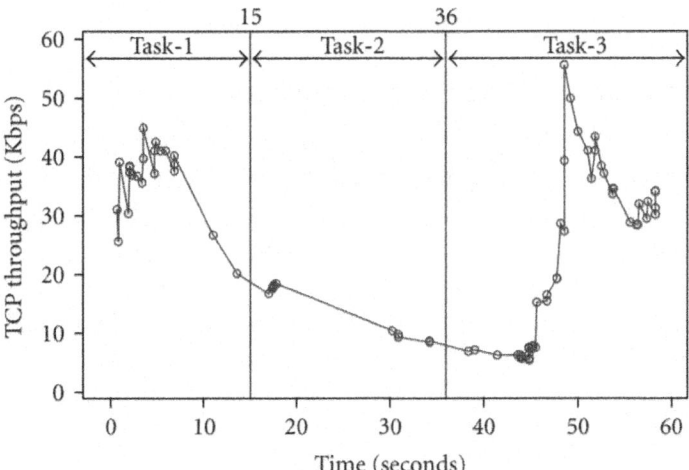

Figure 10: Control traffic (b_{in}) during an expert session on direct GigE network connection.

Network Connection and user Control

Next, we characterize how the network connection quality impacts the trends of user control behavior (i.e., bin). Figures 10 and 11 show the instantaneous bin throughput levels during an expert session on Direct GigE and Public LAN connections, respectively. The throughput levels clearly show the amount of user effort required for accomplishing each of the three tasks of the session. Another notable observation is that user effort is considerably less ($\lceil b_{in} \rceil \approx 60\,Kbps$) in the case of the Direct GigE connection as compared to the user effort ($\lceil b_{in} \rceil \approx 1400\,Kbps$) on the Public LAN connection. Also, the throughput trends are significantly less dense in case of the Direct GigE connection as compared to the Public LAN connection. Due to space constraints, we do not show the throughputs for the WAN network connection, where the expert user effort was the most when compared to the other connections ($\lceil b_{in} \rceil \approx 2000\,\text{Kbps}$).

We note that such an inverse relationship between the network connection quality and user control effort is a driver for the "congestion begets more congestion" phenomenon, where a user expends more effort (i.e., mouse moves/clicks and keyboard strokes) on poor network connections, which cumulatively adds to the congestion already inherent in the poor network connections. The nature of the "Unstable" and "Breakdown" states and their transitions explained in Section 3.2 can be attributed to the occurrence of this particular phenomenon with different intensity levels. The intensity levels are based on the instantaneous network connection quality and the impulsive user reactions to video signal impairments such as frame freezing.

User Behavior and Video image tranSfers

Lastly, we analyze how a user's control behavior and network conditions impact the video image transfers from the microscope at the user end (i.e., b'_{out}). Figure 12 shows the b'_{out} comparison between the novice and expert for varying ψnet conditions. Cross-referring to the q_{mos} equals 5 results shown in Figure 8 for the direct GigE network connection, we can observe that obtaining an "at-the-microscope" user QoE requires end-to-end available bandwidth in excess of 30 Mbps between the user and the microscope ends.

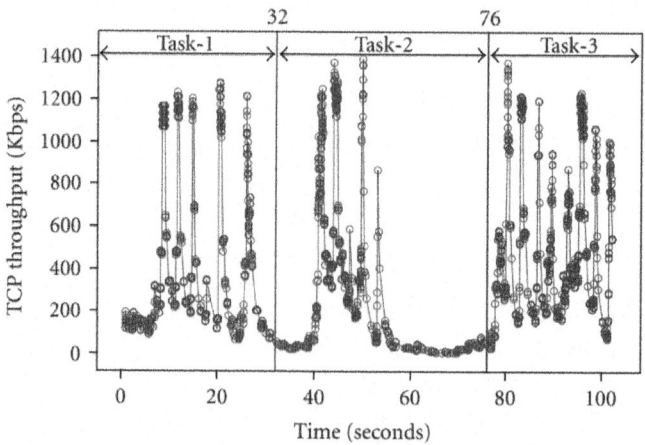

Figure 11: Control traffic (b_{in}) during an expert session on public LAN network connection.

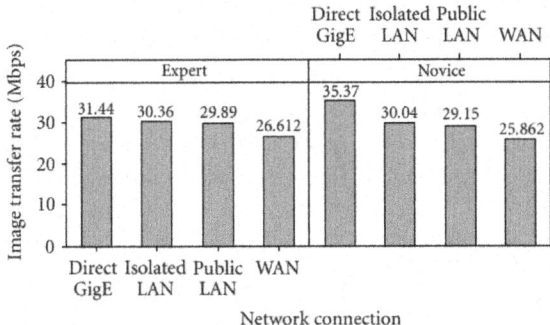

Figure 12: Image transfer rate (b'_{out}) comparison for varying ψ_{net} conditions.

However, it is important to note that the average image transfer rates in remote microscopy can vary based on the microscope functions and user activity. Thus, they may not always be in the range of 30 Mbps. The reason for the high bandwidth consumption in our experiments can be attributed to the KVMoIP solution nature, and activity level in the experiments that had the quad-video panel images in the GUI application shown in Figure 2(a). Such a nature of video may not be present in every user session. For example, there may be sessions whose activity level may be similar to that of Figure 2(b), where the user is mainly plotting graphs, editing parameters while analyzing a sample. For such a session, the end-to-end available bandwidth requirement to achieve "at-the-microscope" user QoE will be considerably less.

REMOTE INSTRUMENTATION COLLABORATION ENVIRONMENT (RICE)

In this section, we describe the Remote Instrumentation Collaboration Environment (RICE) software application we have developed. The RICE design leverages our user and network interplay studies to effectively support the remote observation and operation use-cases for instructors and researchers to train students and/or conduct research from remote locations on the Internet. RICE is based on the UltraVNC solution [10] but has several enhancements targeted for network-aware and collaborative remote instrumentation sessions that are reliable and efficient. The enhancements allow tuning of image feeds based on last-mile network bandwidth, and limit the provision of excessive control given by off-the-shelf VNC solutions during network congestion periods. Also, the enhancements provide collaboration tools (VoIP, chat, presence, control-lock passing) to orchestrate instrument-control amongst multiple remote users during remote operation of expensive and potentially dangerous instruments. Note that the network-aware control blocking and collaboration tools in RICE are features not available in off-the-shelf VNC solutions. The modular software design used in RICE permits customization to cater to unique considerations and requirements of a variety of users and instruments. Thus, RICE is a self-contained collaboration environment for multiple users to reliably and efficiently participate in remote instrumentation sessions.

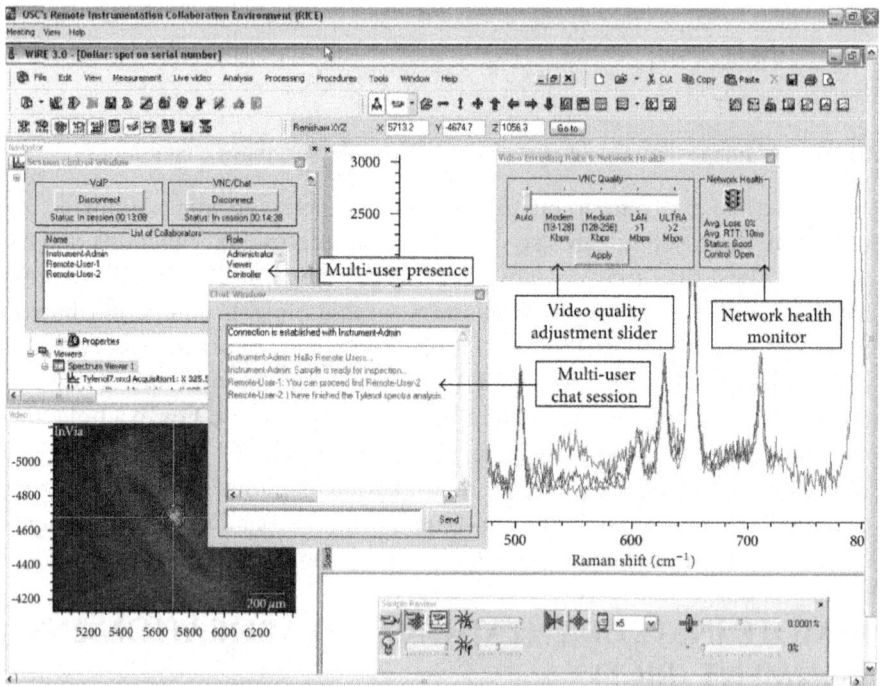

Figure 13: RICE in an active session.

Figure 13 shows the main functionality of RICE in an active session during remote operation of a Renishaw Raman Spectroscope deployed at the Department of Chemistry, The Ohio State University. The "video quality adjustment slider" is used by a RICE client user to manually adjust frame rates and video encoding rates based on the lastmile network bandwidth between the remote user site and instrument lab. A "network health monitor" shows real-time network health in terms of average round-trip delay (RTT) and loss. A traffic light indicates the network health grade as Good (green), Acceptable (amber), and Poor (red). The Good grade corresponds to RTT values in the range [0–150) milliseconds and loss values in the range [0–0.5)%; Acceptable grade corresponds to RTT values in the range (150–300) milliseconds and loss values in the range (0.5–1.5)%; Poor grade corresponds to RTT values in the range (>300] milliseconds and loss values in the range (>1.5]%. Such grade levels of RTT and loss have been obtained by studies such as [22, 23] that have conducted empirical experiments on the Internet for real-time multimedia applications. The network health monitoring in RICE is coupled with network performance anomaly detection using a "plateau-detector algorithm" that warns and blocks user's control-actions during impending and extreme network congestion periods, respectively. The control blocking is essential in spite of the fact that "limit switches" in some well-designed instruments mitigate damage due to user error. This is because it is practically infeasible to take into account all the possible user error cases when designing limit switches. The control status in RICE can be either "open," "warning," or "blocked" depending on the network congestion levels. Details of the network health monitoring coupled with the plateaudetector algorithm implementation in RICE are described in Section 5.1.

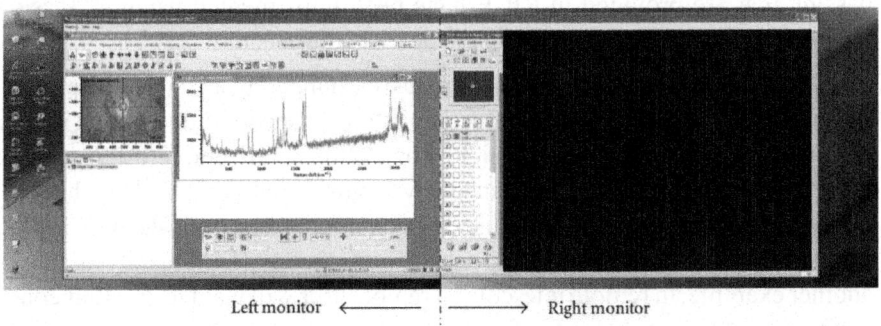

Left monitor ←——— ┆ ———→ Right monitor

Figure 14: Problem with remote dual-screen resolution with UltraVNC.

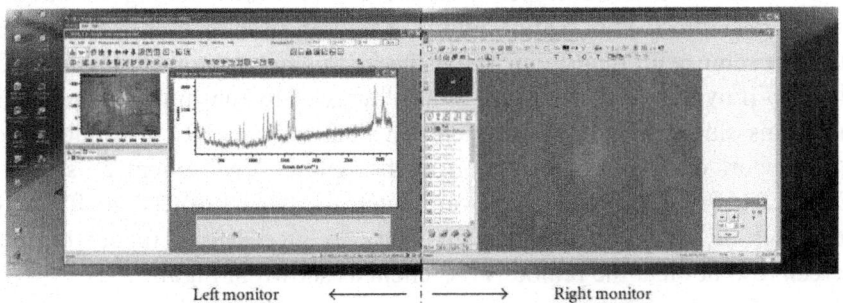

Left monitor ←——— ┆ ———→ Right monitor

Figure 15: Increased remote dual-screen resolution with RICE

RICE is designed to allow multiple remote users to simultaneously connect using RICE clients to the instrument console, which runs an instance of the

RICE server. In a multiuser remote instrumentation session, RICE provides VoIP and chat functionality for collaboration as shown in Figure 13. The VoIP functionality is supported via the open-source OpenMCU software [24] that is integrated within RICE. The chat is supported by a custom chat application integrated within RICE that features colored text that can distinguish messages of the local user from the remote users. The names and roles of the multiple-users connected in a session are displayed and updated in real-time via a presence functionality. The roles correspond to the "administrator," "viewer," or "controller" (i.e., the users who possess the instrument control-lock, and the users who are observers). By default, the administrator always has control privileges, and he/she has the ability to pass another session control-lock to any one of the remote users. Once a remote user has a control-lock, he/she in turn can directly pass that control-lock in a session to any other remote user via the RICE client interface, without intervention of the administrator. All of the above multiuser collaboration tools of VoIP, chat, presence, and control-lock passing that are provided in RICE have been implemented using a "session-signaling protocol," whose details are described in Section 5.2.

RICE has a feature for simultaneously connecting to and transparently switching between two PCs for accomplishing inspection and analysis tasks of a remote instrumentation session. For example, in remote microscopy, a remote user will want to simultaneously connect and switch views between two PCs. One PC corresponds to the electron microscope console and the other PC corresponds to an energy dispersive spectroscopy (EDS) analysis PC. As another example, in remote telescopy, a remote user will want to simultaneously connect to a PC that shows the focal-plane instrument controls and another PC that is performing environment monitoring.

Further, we also developed a feature in RICE for handling remote dual-screen resolution. This feature allows a remote user to mimic an extended desktop setup with dual-monitors at the instrument computer. The extended desktop provides additional real-estate for users to run multiple application programs simultaneously. The openly-available UltraVNC distribution does not support dual-screen resolution at the remote VNC client as shown in Figure 14. We developed a software patch for the UltraVNC in RICE that increases the display resolution geometry to successfully render the dual-screen resolution at the remote VNC client as shown in Figure 15.

In the following subsections, we first describe the plateau-detector algorithm implementation and its integration in network health monitoring in RICE. Thereafter, we describe the session signaling protocol implementation to provide the collaboration tools functionality in RICE.

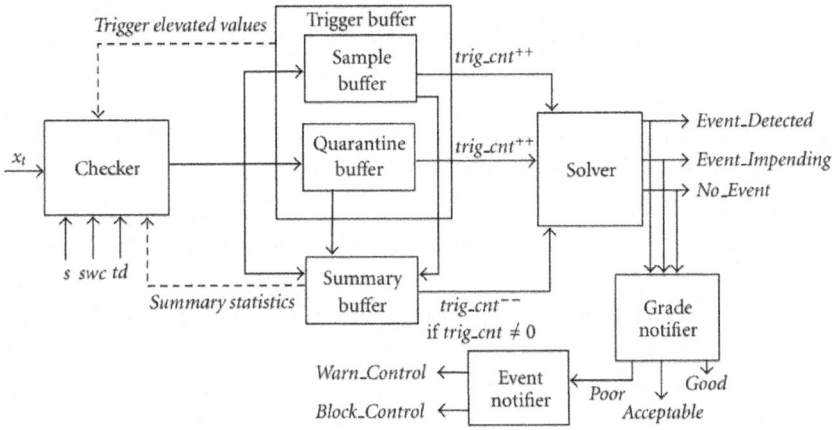

Figure 16: Plateau-detector block diagram.

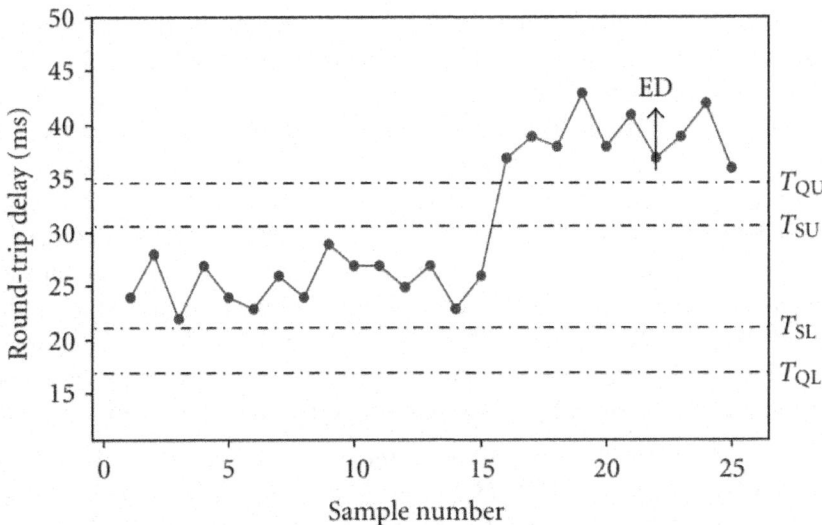

Figure 17: Plateau-detector thresholds illustration.

Network Health Monitoring

We use the popular Ping tool to monitor the network health in a RICE session in terms of average RTT and loss measurements. These RTT and loss metrics are useful to assess the impact of network health on the image transfer performance and interaction responsiveness as perceived by the user. Using RTT and loss measurements of Ping can generally indicate network congestion scenarios in most cases. However, it is possible in some cases that Ping's RTT measurements could misrepresent the network health status. For example, some routers are configured to handle Ping's ICMP packets differently to mitigate denial of service (DoS) attacks. In such cases, the RTT measurements might indicate higher values than actual, which can be construed as due to network congestion. Although there are several other sophisticated network measurement tools such as Iperf [25] and Pathchar [26] that can accurately measure available/per-hop bandwidth, they are not suitable for online network-aware adaptation in RICE. The reason is that these tools consume significant amounts of network bandwidth and CPU resources, and thus interfere with the RICE application performance. Besides, these sophisticated network measurement tools have been designed to be primarily used on network backbones for troubleshooting purposes rather than adaptation purposes in actual applications.

Our sampling rate of RTT and loss between the RICE client and server is periodic with a period of 6 seconds between consecutive samples. Each sample uses the default Ping settings (i.e., four ICMP packets each with packet size of 32 bytes). This results in time-series of instantaneous measurements of RTT and loss. We use xt to denote an instantaneous measurement of RTT or loss that is input to a plateau-detector algorithm that detects network performance anomalies in real time. The purpose of the plateau-detector algorithm is to minimize the false-positive and false-negative anomaly alerts or triggers that arise if a naive mean-based method is used. A false-positive trigger is one that gets reported when there is no actual anomaly, whereas a false-negative trigger is one that does not get reported when in fact there is an actual anomaly. Two instances of plateau-detector algorithm run in-band in every RICE session, one to detect RTT anomalies, and the other to detect loss anomalies. The overall network health grade g and control status c are determined using an OR function of the two plateau-detector algorithm instance outputs.

Our plateau-detector algorithm implementation is similar to the basic implementations found in [27, 28] that are used for routine network monitoring. However, our implementation has several modifications to suit the time scales and user reaction times during network congestion periods in RICE sessions. In the mean-based method, the network health norm is determined by calculating the mean μ and comparing it with the standard deviation σ for a set of x_t values sampled most recently into a summary buffer. The number of samples in the summary buffer is user-defined and is specified using a summary window count swc. In comparison to the mean-based method, the plateau-detector requires two additional user-defined inputs called trigger duration td and sensitivity s. The trigger duration td specifies the duration of the anomaly before a trigger is signaled. Thus, the smaller the td, the faster a trigger will be signaled in the event of an anomaly. However, the td must be chosen to be large enough such that the transient spikes or bursts (i.e., noise events in network health are not signaled). The sensitivity s specifies the magnitude of the change for it to be considered as an anomaly. The choice of the s again requires consideration of the tradeoffs (i.e., a small s results in triggers for slight variations in network performance magnitudes, whereas a large s could overlook actual anomalies that should be detected).

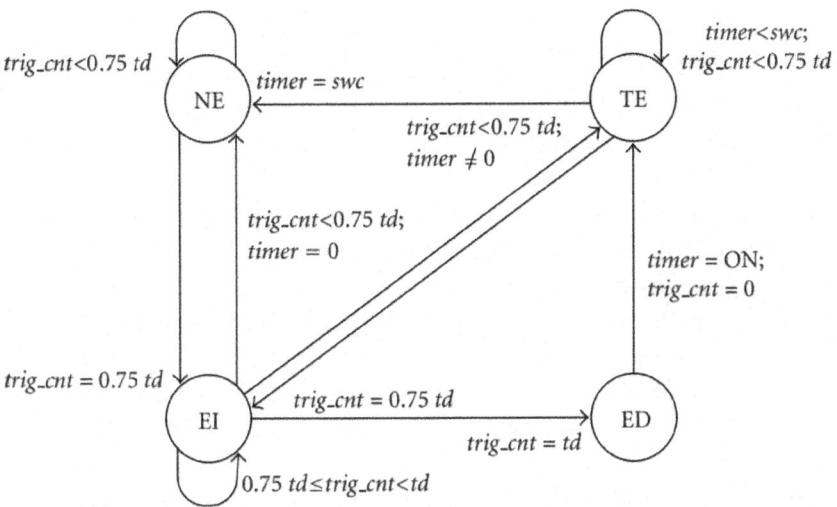

Figure 18: Plateau detector states and state-transitions.

Given our network health sampling rate of 6 seconds in RICE, we choose the swc to be equal to 20 so that anomalies can start getting detected within 2 minutes of initiation of a RICE session. In accordance with the td and s values selection in [27, 28], we choose td to be approximately 1/3rd swc (i.e., 7), and s value is chosen to be equal to 2. In this context, we remark that we have validated our selection of s = 2 using extensive simulations on synthetic and actual network health time series. The validation results are beyond the scope of this paper. Our general observation has been that values of s in the range of 2 produce the least number of false triggers (sum of false positives and false negatives), for a wide selection of swc and td values.

Figure 16 shows the different components of our plateau detector algorithm implementation. The values of x_t are first input to a "checker" which compares whether the most recent x_t value lies within the upper and lower thresholds of the (i) summary buffer sumbuff (i.e., T_{SU} and TSL) or (ii) quarantine buffer qbuff (i.e., T_{QU} and T_{QL}). These thresholds are illustrated in Figure 17 and are calculated using mean μ and standard deviation σ of the summary window as follows:

$$T_{SU} = \mu + s * \sigma,$$

$$T_{QU} = \mu + 2 * s * \sigma,$$

$$T_{SL} = \mu - s * \sigma,$$

$$T_{QL} = \mu - 2 * s * \sigma.$$

(8)

If x_t values lie within these thresholds, the plateaudetector will be in the no event (NE) state shown in Figure 18. In this state, x_t values are put into the sumbuff. If x_t values go below T_{QL} or exceed T_{QU}, they are put into the quarantine buffer qbuff. Similarly, if x_t values cross TSL and TSU , they are put into the sample buffer sampbuff. If x_t is put into either qbuff or sampbuff, trigger count trig cnt is incremented. Whereas, if x_t is put into sumbuff, trig cnt is decremented as long as trig cnt is nonzero. If trig cnt exceeds 0.75*td due to increasing number of x_t values going into qbuff or sampbuff, then the plateau-detector enters into an event impending (EI) state. If the trig cnt drops below 0.75*td, then the plateau-detector returns to NE state. Otherwise, the plateau-detector stays in the EI state until trig cnt equals td, after which it enters into an event detected (ED) state. Figure 17 shows an event detection occuring after x_t crosses the thresholds for the td of 7 samples. At this point, the trig cnt is reset, and a timer is turned ON. The plateaudetector now goes into a trigger elevated (TE) state, where the upper and lower thresholds are calculated as follows:

$$T'_{SU} = 1.2 * \max(x_t) \text{ in } trigbuff,$$

$$T'_{QU} = 1.4 * \max(x_t) \text{ in } trigbuff,$$

$$T'_{SL} = 0.8 * \max(x_t) \text{ in } trigbuff,$$

$$T'_{QL} = 0.6 * \max(x_t) \text{ in } trigbuff.$$

(9)

Until the timer equals swc, the elevated thresholds are used for comparing x_t. The reason for the trigger elevation is to avoid reporting of repeated triggers for the already detected anomaly. It is relevant to note that the plateaudetector can transition from TE state to EI state if another

(1) **Input:** summary window count (swc), sensitivity (s), trigger duration (td), instantaneous measurement (x_t)
(2) **Output:** mean μ, network health grade g, control status c
(3) **begin procedure**
(4) **repeat**
(5)   **for** each new x_t **do**
(6) $g = No\_Event; c = Allow\_Control$
(7)   Calculate mean (μ) and standard deviation (σ) of summary window buffer ($sumbuff$)
(8)   **if** (timer == 0 /* NE state */) **then**
(9)   Calculate upper (T_{SU}, T_{QU}) and lower (T_{SL}, T_{QL}) thresholds using μ and σ of $sumbuff$ and s
(10)   **else**
(11)   /* TE state */ Calculate upper (T_{SU}, T_{QU}) and lower (T_{SL}, T_{QL}) elevated thresholds using $trigbuff$
(12)   **end if**
(13)   /* Compare x_t with the thresholds and assign x_t to appropriate buffer */
(14)   **if** ($T_{QL} > x_t$ or $x_t > T_{QU}$) **then**
(15)   Put x_t into quarantine buffer $qbuff$; Increment $trig\_cnt$
(16)   **end if**
(17)   **if** ($T_{SL} > x_t$ or $x_t > T_{SU}$) **then**
(18)   Put x_t into sample buffer $sampbuff$; Increment $trig\_cnt$
(19)   **end if**
(20)   /* Report anomaly event types */
(21)   **if** ($trig\_cnt > td$) **then**
(22) $event\_type = Event\_Detected$; Copy $sampbuff$ and $qbuff$ to $sumbuff$; Reset $trig\_cnt$
(23)   **else if** ($trig\_cnt > 0.75^* td$) **then**
(24) $event\_type = Event\_Impending$
(25)   **else**
(26) $event\_type = No\_Event$
(27)   **end if**
(28)   /* Update the buffers with latest network health norm */
(29)   **if** ($trig\_cnt == 0$ and $trigbuff$ not empty) **then**
(30)   Copy $sampbuff$ to $sumbuff$; Empty $qbuff$
(31)   **end if**
(32)   /* Report μ of $sumbuff$, g and c */
(33)   **if** (magnitude of x_t in $Good$ grade level) **then**
(34)   return μ, $g = Good$, $c = open$
(35)   **else if** (magnitude of x_t in $Acceptable$ grade level) **then**
(36)   return μ, $g = Acceptable$, $c = open$
(37)   **else if** (magnitude of x_t in $Poor$ grade level) **then**
(38)   **if** ($event\_type == Event\_Detected$) **then**
(39)   return μ, $g = Poor$, $c = Block\_Control$
(40)   **end if**
(41)   **if** ($event\_type == Event\_Impending$) **then**
(42)   return μ, $g = Poor$, $c = Warn\_Control$
(43)   **end if**
(44)   **end if**
(45) **end for**
(46) **until** end of RICE session
(47) **end procedure**

Algorithm 1: Plateau-detector algorithm implementation in RICE.

Collaboration Tools

Herein, we describe the session signaling protocol we developed to provide the collaboration tools functionality in RICE. The protocol involves exchanging

TCP-based messages on port 6000 between the RICE clients of the remote users and the RICE server at the instrument console. Figures 19(a) and 19(b) show the start session and end session messages, respectively, for a RICE session involving two remote users. To start a session, a RICE client sends a START_ REQ message to the RICE server. If a RICE server accepts the connection, it sends a START_ACK message back to the RICE client. Similarly, if a RICE client wants to terminate a session, it sends an END_REQ message to the RICE server, which in turn sends an END_ACK message to the RICE client. The UPDATE_LIST message maintains the latest presence information of all the users. It is event-driven and is sent from the RICE server to each of the RICE clients whenever there is a change of status in the RICE clients (i.e., whenever RICE clients join/leave the session, and whenever RICE clients switch between "viewer" and "controller" roles).

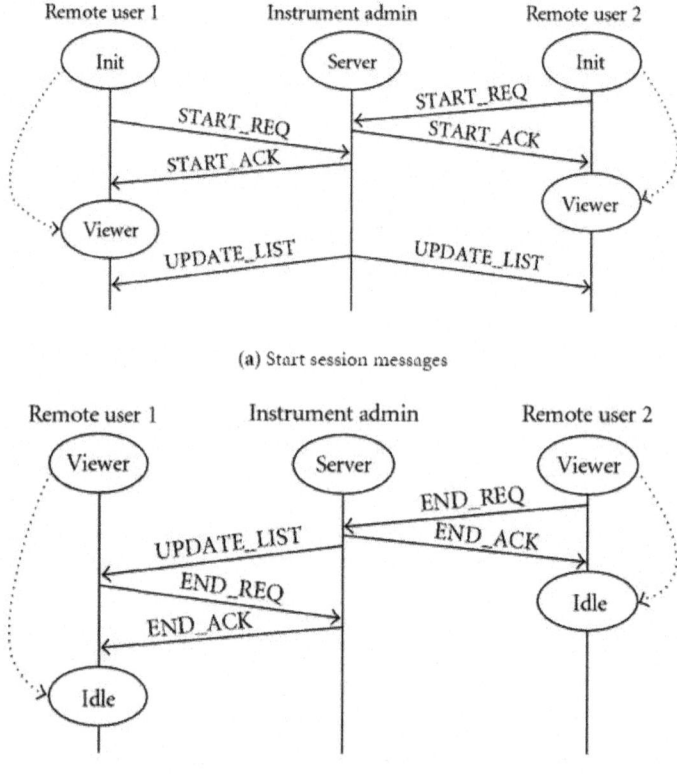

(a) Start session messages

(b) End session messages

Figure 19: RICE's Session Signaling Protocol messages for session initiation and termination.

Figure 20 shows the different messages exchanged during a session when the collaboration tools are being used by the remote users. For the control-lock passing, we require the instrument administrator to initially pass the control-lock to any remote user. For this, the PASS_LOCK message is used. As we noted earlier, the instrument administrator always has a copy of the control-lock, and can retrieve the other control-lock from a remote user at any time. For the control-lock retrieval, the GET_LOCK message is used. Once a control-lock is given to a remote user, we also support the direct control-lock passing to another remote user, without the intervention of the instrument administrator. For this, the RICE client that has the control-lock sends a PASS_LOCK_HOST to the RICE server, which in turn sends a PASS_LOCK to the appropriate RICE client instantly. A similar relaying of chat text between remote users is performed by the RICE server. Here, the chat text from a RICE client is sent to the RICE server using the CHAT_TEXT message, which in turn sends an UPDATE_CHAT_TEXT message to all the other RICE clients. We remark that the relaying of VoIP traffic between the RICE clients is handled by the OpenMCU software using the standard ITU-T H.323 signaling protocols.

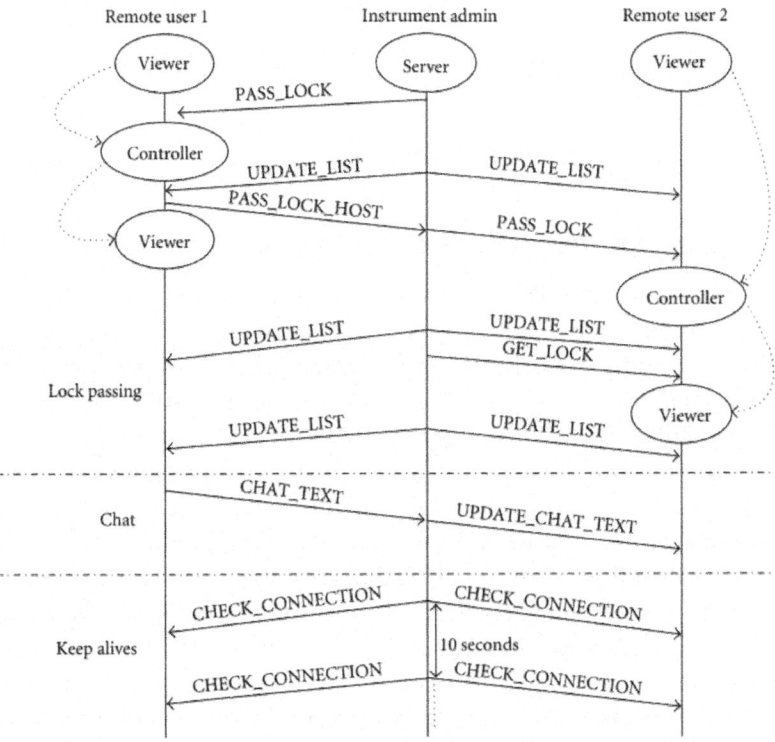

Figure 20: RICE's Session Signaling Protocol in-session messages.

In case of any exception conditions, we use a time-out period of 10 seconds. Exception conditions correspond to cases where the RICE server is not running/accessible, or if one or more RICE clients are disconnected due to a network partition during an active session. To detect the latter case, CHECK_ CONNECTION messages are sent every 10 seconds from the RICE server to each of the RICE clients. Table 1 summarizes the purpose and format of the various session signaling protocol messages in RICE.

Table 1: List of RICE's session signaling protocol messages

Message name	Purpose	Format
START_REQ	Sent from RICE client to RICE server to request a connection establishment	(START_REQ, USER_NAME, HOST_IP)
START_ACK	Sent from RICE server to RICE client to accept a connection request	(START_ACK, ADMIN_NAME)
UPDATE_LIST	Sent from RICE server to all connected RICE clients to update changes in user names/status	(UPDATE_LIST, USERS_STATUS[][])
PASS_LOCK	Sent from RICE server to a RICE client to grant instrument control privileges	(PASS_LOCK)
PASS_LOCK_HOST	Sent from a RICE client to RICE server to pass control privileges to another RICE client	(PASS_LOCK, HOST_IP)
GET_LOCK	Sent from RICE server to a RICE client to revoke instrument control privileges	(GET_LOCK)
CHAT_TEXT	Sent from a RICE client to RICE server to send a chat text message to other connected RICE clients	(CHAT_TEXT, TEXT_INFO)
UPDATE_CHAT_TEXT	Sent from RICE server to all connected RICE clients to relay a new chat text message	(UPDATE_CHAT_TEXT, TEXT_INFO)
CHECK_CONNECTION	Sent from RICE server to all connected RICE clients to check connection validity	(CHECK_CONNECTION)
END_REQ	Sent from RICE client or RICE server to terminate a connection	(END_REQ)
END_ACK	Sent from RICE client or RICE server to acknowledge termination of a connection	(END_ACK)

Applications for Research and Training

In this section, we describe a few applications of RICE to foster research and training activities involving remote instrumentation. In the context of research, RICE allows multiple experts to jointly collaborate in investigations of samples loaded in scientific instruments. Each of the remote experts can use the quality adjustment slider to easily configure the appropriate frame rates of the image feeds from the instrument to match their respective last-mile network links. During a remote instrumentation session, the self-contained collaboration tools such as VoIP and chat in RICE enable the experts to efficiently communicate with each other. The control-lock passing feature ensures that there are no conflicts in remote operation of the instrument amongst the experts. Using RICE, each expert after completing his/her set of tasks on the instrument can instantly transfer the instrument-control to any other expert who wants to perform another set of tasks on the instrument. Both the control-lock passing and network health monitoring in RICE prevent inadvertent damages to the instrument by the remote users.

In the context of training, all of the above benefits of RICE mentioned for research are equally applicable, except the roles of remote users which are different. A number of students can use RICE to join and participate in a remote instrumentation session that is controlled by an instructor, who also could be remote using RICE. After the training session, the instructor can assign time-slots on the instrument for the students to complete their lab assignments. During the beginning and end of the lab time-slots of the students, the local instrument administrator can pass and retrieve the control-lock amongst the students, respectively.

RICE can be customized for integrating remote instrumentation with scientific web-portals that help in organization and archival of images and datasets acquired from instruments. Specifically, web-services and middleware defined in works such as [9, 29] can be implemented within RICE to automatically upload images and datasets to the web-portal storage along with basic metadata such as instrument type and date and time stamps. Remote users can then login to the web-portal using their RICE account credentials and supply additional tags and annotations to their respective image and datasets for search and archival functions. If necessary, the users can even submit batch jobs of image processing filters or data analytics to compute clusters via the same web-portal interface. Thus, RICE can be an essential component in cyberinfrastructure deployments that aim at integrating instruments with networking, computing and storage resources for catering to the research and training needs of scientific user-communities.

CONCLUSION

In this paper, we modeled and characterized the complex interplay between the user control behavior and video image transfer performance in remote instrumentation sessions. Our remote instrumentation session model borrowed demand and supply terminology from traditional economics and identified the various system states (Idle, Stable, Unstable, Breakdown, and Recovery) and their transition conditions. The transition conditions were found to be primarily driven by time-varying user behavior and connection quality of the network path between the user and the instrument. Analyzing subjective and objective measurements of remote microscopy sessions involving actual users on LAN and WAN network paths, we found that (a) user QoE is highly sensitive to network health fluctuations caused by network congestion, (b) network health impacts both the user's control traffic throughput and microscope's video image transfer rates, and network congestion can hamper user productivity and cause the system to enter into unproductive states (i.e., Unstable and Breakdown states), and (c) the real-time control and video image transfer traffic is extremely bandwidth intensive for achieving "at-the-microscope" QoE. Thus, we demonstrated that remote instrumentation is a demanding network-based immersive multimedia application and is comparable to other applications of its class such as online-gaming, and high-definition videoconferencing.

We also described our RICE software functionalities that leverage our user and network interplay studies to cope with network bottlenecks, and cater to the requirements of remote observation and remote operation use-cases. In particular, we described in detail two main RICE functionalities that facilitated reliable and efficient remote instrumentation. The first functionality corresponded to the network health monitoring coupled with network performance anomaly detection using a "plateau-detector algorithm" that warns and blocks user's control-actions during network congestion periods. The second functionality corresponded to the "session-signaling protocol" we developed to enable multiuser collaboration in RICE using VoIP, chat, presence, and control-lock passing. Finally, we presented potential applications of RICE for research and training purposes that require remote instrumentation capabilities.

ACKNOWLEDGMENTS

This work has been supported in part by The Ohio Board of Regents and The Ohio State University's CAMM VIM program. The authors thank Dr. Peter Collins, Daniel Huber, Robert Williams, and Professor Hamish Fraser of OSU's CAMM for their participation in the remote microscopy testbed, RICE development feedback, and for providing microscope sample graphics used in this paper. They also thank Gordon Renkes (Department of Chemistry, The Ohio State University) for RICE testing, development feedback, and screenshots of RICE in active sessions. Further, they thank Dr. Dave Hudak, Neil Ludban, Dr. Eylem Ekici, Dr. Dong Xuan, and Dr. Steve Gordon for their useful suggestions during the course of the study. A preliminary version of this paper has appeared in the proceedings of ICST/ACM Immersive Telecommunications Conference (IMMERSCOM), 2007 [30].

REFERENCES

1.   "Chemistry Research Instrumentation and Facilities: Departmental Multi-User Instrumentation (CRIF:MU)," Program Solicitation NSF 07-552, National Science Foundation, Arlington, Va, USA, 2007.

2.   T. Richardson, Q. Stafford-Fraser, K. R. Wood, and A. Hopper, "Virtual network computing," IEEE Internet Computing, vol. 2, no. 1, pp. 33–38, 1998. ·

3.	K. W. Hodapp, J. B. Jensen, E. M. Irwin, et al., "The Gemini Near-Infrared Imager (NIRI)," Publications of the Astronomical Society of the Pacific, vol. 115, no. 814, pp. 1388–1406, 2003. ·

4.	K. Jeffay, T. Hudson, and M. Parris, "Beyond audio and video: multimedia networking support for distributed, immersive virtual environments," in Proceedings of the 27th Euromicro Conference (EUROMICRO '01), pp. 300–307, Warsaw, Poland, September 2001. ·

5.	K. Furuya, M. Tanaka, K. Mitsuishi, et al., "Public opened internet electron microscopy in education field," in Proceedings of the Microscopy and Microanalysis Conference, vol. 10, pp. 1566–1567, Savannah, Ga, USA, August 2004.

6.	J. Kao, D. Troxel, and S. Kittipiyakul, "Internet remote microscope," in Telemanipulator and Telepresence Technologies III, vol. 2901 of Proceedings of SPIE, pp. 90–100, Boston, Mass, USA, November 1996. ·

7.	M. A. O'Keefe, B. Parvin, D. Owen, et al., "Automation for on-line remote-control in-situ electron microscopy," in Proceedings of the Pfefferkorn Conference on Electron Image and Signal Processing, Silver Bay, NY, USA, May 1996.

8.	M. Hadida, Y. Kadobayashi, S. Lamont, et al., "Advanced networking for telemicroscopy," in Proceedings of the of the 10th Annual Internet Society Conference on Advanced Networking for Telemicroscopy (INET '00), Yokohama, Japan, July 2000.

9.	T. E. Molina, G. Yang, A. W. Lin, S. T. Peltier, and M. H. Ellisman, "A generalized service-oriented architecture for remote control of scientific imaging instruments," in Proceedings of the 1st International Conference on e-Science and Grid Computing, vol. 2005, pp. 550–556, Melbourne, Australia, December 2005. ·

10.	Ultra VNC Open-source Remote Access Solution, http://www.uvnc.com/.

11.	Real VNC Open-source Remote Access Solution, http://www.realvnc.com/.

12.	Thinklogical KVMoIP Remote Access Solution, http://www.thinklogical.com/.

13.	Avocent KVMoIP Remote Access Solution, http://www.avocent.com/.

14.	Adder KVMoIP Remote Access Solution, http://www.adder.com/.

15.	M. C. Wright, C. R. Hubbard, R. Lenarduzzi, and J. Rome, "Internet-based remote collaboration at the neutron residual stress facility at HFIR," in Proceedings of the Workshop on New Opportunities for Better User Group Software (NOBUGS '02), Gaithersburg, Md, USA, November 2002.

16. R. H. Cockrum, D. L. Clark, and S. T. Kelly, "Remote internet instrumentation for monitoring ocean data," in Proceedings of the MTS/IEEE Conference and Exhibition (OCEANS '01), vol. 2, pp. 1176–1182, Honolulu, Hawaii, USA, November 2001. ·

17. The Ohio State University CAMM VIM Program, http://www.camm.ohio-state.edu/.

18. S. Winkler, Digital Video Quality: Vision Models and Metrics, John Wiley & Sons, New York, NY, USA, 2005.

19. P. Calyam, D. Krymskiy, M. Sridharan, and P. Schopis, "TBI: end-to-end network performance measurement testbed for empirical bottleneck detection," in Proceddings of the 1st International Conference on Testbeds and Research Infrastructures for the Development of Networks and Communities (Tridentcom '05), pp. 290–298, Trento, Italy, February 2005. ·

20. S. Shalunov and B. Teitelbaum, "One-way Active Measurement Protocol (OWAMP) Requirements," IETF RFC 3763, 2004.

21. J. Mullin, L. Smallwood, A. Watson, and G. Wilson, "New techniques for assessing audio and video quality in real-time interactive communications," ETNA Project Report, IHM-HCI, Lille, France, 2001.

22. ITU-T Recommendation G.114, "One-Way Transmission Time," 1996.

23. P. Calyam, M. Sridharan, W. Mandrawa, and P. Schopis, "Performance measurement and analysis of H. 323 traffic," in Proceedings of the 5th International Workshop on Passive and Active Network Measurement (PAM '04), pp. 137–146, Antibes Juan-les-Pins, France, April 2004.

24. Open H.323 Project, http://sourceforge.net/projects/openh323.

25. A. Tirumala, L. Cottrell, and T. Dunigan, "Measuring end-to-end bandwidth with Iperf using Web100," in Proceedings of the 4th International Workshop on Passive and Active Network Measurement (PAM '03), La Jolla, Calif, USA, April 2003.

26. A. B. Downey, "Using pathchar to estimate Internet link characteristics," in Proceedings of the ACM SIGMETRICS International Conference on Measurement and Modeling of Computer Systems (SIGMETRICS '99), vol. 27, no. 1, pp. 222–223, Atlata, Ga, USA, May 1999. ·

27. A. McGregor and H.-W. Braoun, "Automated event detection for active measurement systems," inProceedings of the Workshop on Passive and

Active Measurements (PAM '01), Amsterdam, The Netherlands, April 2001.

28. C. Logg, L. Cottrell, and J. Navratil, "Experiences in traceroute and available bandwidth change analysis," in Proceedings of the ACM SIGCOMM Workshop on Network Troubleshooting: Research, Theory and Operations Practice Meet Malfunctioning Reality, pp. 247–252, Portland, Ore, USA, August-September 2004. ·

29. I. Atkinson, D. du Boulay, C. Chee, et al., "Developing CIMA-based cyberinfrastructure for remote access to scientific instruments and collaborative e-research," in Proceedings of the 5th Australasian Symposium on ACSW Frontiers, vol. 249 of ACM International Conference Series, pp. 3–10, Ballarat, Australia, January-February 2007.

30. P. Calyam, N. Howes, A. Kalash, and M. Haffner, "User and network interplay in internet telemicroscopy," in Proceedings of the 1st ICST/ ACM International Conference on Immersive Telecommunications (IMMERSCOM '07), Verona, Italy, October 2007.

Chapter 3

EFFICACY OF ELECTRONIC FORAMEN LOCATORS IN CONTROLLING ROOT CANAL WORKING LENGTH DURING ROTARY INSTRUMENTATION

Lorena Arruda Parente[1] , Martin D. Levin[2] , Rodrigo Ricci Vivan[3] , Ricardo Affonso Bernardes[4] , Marco Antonio Hungaro Duarte[3] , Bruno Carvalho de Vasconcelos[1]

[1]School of Dental Medicine of Sobral, UFC - Universidade Federal do Ceará, Sobral, CE, Brazil

[2]Department of Endodontics, School of Dental Medicine, University of Pennsylvania, Philadelphia, PA, USA

[3]Department of Dentistry, Endodontics and Dental Materials, Bauru Dental School, USP - Universidade de São Paulo, Bauru, SP,Brazil

[4]ABO - Brazilian Dental Association, Taguatinga, DF, Brazil

ABSTRACT:

The present study evaluated the efficacy of electronic foramen locators (EFLs) to control root canal working length during rotary instrumentation and to assess possible reliability variations of different working lengths. Forty-eight human mandibular bicuspids were randomly divided in 2 groups according to the used device, Root ZX II (RZX) and Propex II (PRO). They were further subdivided in 2 subgroups according to the root canal preparation level (0.0 and -1.0). Preparation was performed with the Protaper rotary system using a crown-down technique. RZX was employed on its automatic auto-reverse mode (AAR) and PRO was used with the MPAS-10R contra-angle to monitor the preparation. The last used file (F3) was fixed, and the apical portion of the teeth was worn buccolingually, allowing to measure the extent between the file tip and the apical foramen (AF). The precision values of 0.0 mm and -1.0 mm were 100% and 0.0% for RZX, and 100% and 66.7% for PRO, respectively, with a range of ±0.5 mm. Statistical analysis showed no differences between the groups at 0.0 mm. However, at -1.0 mm, RZX showed the poorest results (0.96±0.11 mm), followed by PRO (0.43±0.23 mm). The difference between RZX and PRO was statistically significant. The EFLs were precise in maintaining the working length during rotary preparation when reaching the AF, but when their penetration was limited, both devices showed decreased precision; the RZX AAR failed in all instances.

INTRODUCTION

Precisely determining root canal length is one of the key steps leading to successful outcomes and treatment safety ([1] [2] [3]). Studies have shown that electronic foramen locators (EFLs) are extremely effective, reaching success rates greater than 80% *ex vivo* ([4] [5] [6] [7]) or *in vivo* ([2] [8] [9] [10] [11] [12]).

Root ZX (J. Morita, Tokyo, Japan) works by determining root canal length by calculating the impedances at two frequencies, 0.4 and 8.0 kHz, measured simultaneously [2] [5] [9] [13]. Studies have shown that Root ZX presented reliable results, providing accurate values over 90% when used even under unfavorable conditions ([5] [12] [14] [15] [16] [17]).

Recently, hand pieces with integrated EFLs have been employed to assist in the mechanical preparation of the root canal system. This arrangement has increased use because of its simplicity, improved workflow and reliability ([1] [2] [3] [7] [18]). One such device is Root ZX II (J. Morita, Tokyo, Japan), which associates 2 modules that allow the integration of an EFL with an electric motor for mechanical preparation ([1] [2] [3] [7] [18]). Another possibility is to integrate an EFL and a low-speed MPAS-10R contra-angle (NSK, Tokyo, Japan) into a single unit, further simplifying the equipment. It combines with an electric motor (X-Smart; Dentsply-Maillefer, Ballaigues, Switzerland), allowing the operator constant monitoring of the position inside the root canal system [2]. This device could be coupled to any EFL. One of them, the Propex II (Dentsply-Maillefer), is based on the determination of the square root of the mean impedances at two frequencies, 0.5 and 8.0 kHz, measured separately and compared to the reference values of the device memory. According to the manufacturer, Propex II has the advantage of improved accuracy due to reduced electronic noise [15]. Studies evaluating its accuracy in *ex vivo* and *in vivo* conditions with manual instruments showed precision values up to 90% [12] [15] [17].

The apical limit of root canal instrumentation should extend to 1.0 mm short of the radiographic apex, a position that would represent the *loci* of the apical constrictions [19] [20]. However, based on instrument rigidity, apical curvature and anatomy of the apical foramen (AF), some researchers suggest different instrumentation end-points. The main benefits to extend preparation up to the

AF include reducing the risks of apical deviations and eliminating bacterial contamination possibly located in the last 1.0 mm of the root canal [16][21][22]. However, the use of the hybrid devices limited to positions short of the AF may compromise the accuracy and reliability of the EFLs [7][9][17].

It is well established that the combination of EFLs with slow speed hand pieces improves workflow, however the accuracy of these strategies (e.g., Root ZX II and Propex II/MPAS-10R) has been questioned [2][3]. Accordingly, the aim of this study was to evaluate the precision of these hybrid devices and possible variations caused by different working lengths (0.0 mm and -1.0 mm short of the AF).

MATERIAL AND METHODS

Forty-eight human single-rooted mandibular bicuspids with complete root formation, extracted for orthodontic, prosthetic and/or periodontal reasons were selected for this study after approval by the Research Ethics Committee of the Federal University of Ceará (Protocol. #531.206/2014). The selected teeth were limited to premolars with Vertucci Type I roots and patent apical foramina measuring less than 200 µm. Teeth with extensive caries, significant curvatures, root resorption or fractures were excluded from the study.

Access opening was created using a water-cooled high-speed handpiece, diamond burs #1012 and #3081 (KG Sorensen Ind., Barueri, SP, Brazil) in a standard technique. Initial canal exploration was performed with manual #10 K-files (Dentsply-Maillefer) in order to verify the presence of a single root canal and foraminal patency. At this moment the original diameter of the AF was determined adjusting a K-file in its opening observed with a clinical microscope using 40x magnification (Alliance, Campinas, SP, Brazil). Canals in disagreement with the inclusion criteria were substituted. Then, the specimens were numbered and the root canal lengths were determined under magnification. The foramina were then instrumented to a #20 K-file.

Biomechanical preparation of cervical and middle thirds of the root canals was performed with the Protaper System S1 and SX instruments (Dentsply-Maillefer), using the crown-down technique. The instruments were rotated in an electric motor (X-Smart; Dentsply-Maillefer) according to manufacturer's recommendations (300 rpm and 2.0 N) 5.0 mm from the AF. Irrigation was made between each instrument insertion with 1.0 mL of 2.5% sodium hypochlorite (Biodinâmica, Ibiporã, PR, Brazil) with an appropriate irrigating syringe and needles (Navitip; Ultradent, South Jordan, UT, USA). After the preparation of the cervical and middle thirds of the root canal, the excess irrigating solution was suctioned from the pulp chamber keeping the root canal moist and maintaining foraminal patency.

After the preparation of the cervical and middle thirds of the root canal was completed, the teeth were randomly divided in 2 groups according to the used EFL device: Group 1 - Root ZX II; and Group 2 - Propex II/MPAS-10R. These groups were divided into 2 subgroups depending on the extent of the apical preparation (i.e. working length): 0.0 mm (n=12) and 1.0 mm short from the AF (n=12). The teeth were then attached to a support and their root apices were immersed in alginate (Jeltrate II; Dentsply, Petrópolis, RJ, Brazil) to help establish contact with the labial clip of the device. The experiment was performed in groups of 6 or fewer teeth in fresh alginate mixed no more than 30 min before measurements were performed.

A single researcher calibrated and blinded to the previously determined real length, conducted root canal instrumentation of the specimens. The sequence of instrumentation followed the Protaper manufacturer's recommendations with the crown-down technique, concluding with an F3 file. 1.0 mL of 2.5% sodium hypochlorite was used between each file insertion and the foraminal patency was checked. The Root ZX II employed calibration of the automatic auto-reverse (AAR) function based on both apical limits (0.0 mm and 1.0 mm). The Propex II/MPAS-10R device used the apical limit presented by the ELF, as determined by the operator. Following root canal preparation, the last instrument was disconnected from the contra-angle and fixed in place at the assigned working length with cyanoacrylate adhesive (Super Bonder; Loctite of Brazil, São Paulo, SP, Brazil).

The teeth were visualized under the clinical microscope at a 16x magnification and had their apical 4.0 mm carefully worn with a diamond tip (#3082; KG Sorensen) in buccolingual direction to allow for the visualization of the entire apical portion of the root canal. The last layer of dentin was removed with a scalpel blade (#15; Embramac, Campinas, SP, Brazil). The specimens were photographed with a digital camera attached to the microscope at 40x magnification (Fig. 1) and analyzed with Image Tools 3.0 software (UTHSCSA, San Antonio, TX, USA). The photomicrographs were blindly analyzed, with negative and positive values assigned to measurements indicating short and beyond the AF, respectively. Statistical analysis was carried out to measure the differences between the absolute mean errors of devices, measured in millimeters. The Shapiro-Wilk test confirmed the parametric nature of the results; they were submitted to the ANOVA and Bonferroni test with significance set at 5%.

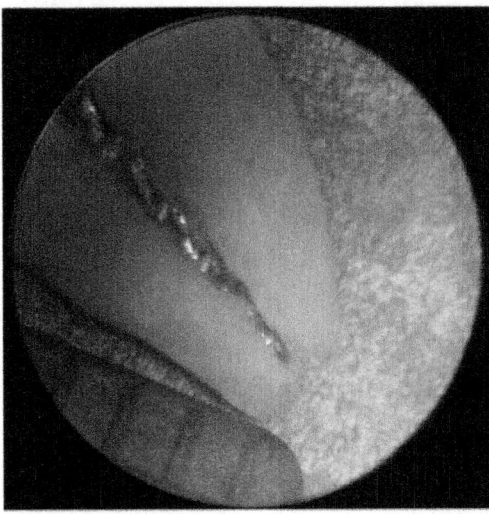

Figure 1: Image captured at 40′ magnification presenting a specimen after the preparation of root apex for the error measurement.

RESULTS

Table 1 presents the average errors with standard deviation, as well as the values found for each group at 0.0 mm and -1.0 mm from the AF. Statistical analysis showed no significant difference between the devices when measurements at the AF were analyzed (0.0 mm) (p>0.05). However, when the penetration of the instruments was set at -1.0 mm, a significant difference was detected between the devices when compared to the other groups (p<0.05).

Table 1: Distance (mm) from device measurements to 0.0 and -1.0

| Device | 0.0 | | | | -1.0 | | | |
| | Mean* | SD | Margin | | Mean* | SD | Margin | |
			Minimum	Maximum			Minimum	Maximum
Root ZX II	0.11a	0.08	-0.26	0.19	0.96c	0.11	0.82	1.11
Propex II + MPAS	0.22a	0.12	-0.33	0.00	0.43b	0.23	-0.40	0.78

*Mean error calculated in terms of absolute values of the determinations. Different superscript letters indicate statistically significant differences according to the one-way ANOVA and Bonferroni tests (p<0.05).

Tables 2 and [3] present the distribution and percentages of each device with respect to the preparations at 0.0 mm and -1.0 mm, respectively. Accuracy of the devices with a tolerance of ±0.5 mm at 0.0 mm was 100% for both the Root ZX II and Propex II. At -1.0 mm, using the same tolerance level, the accuracy of Root ZX II was 0.0% and of Propex II was 66.7%. Instrumentation beyond the apical limit occurred in 33.3% of the Root ZX II instrumented teeth and 0.0% with the Propex II instrumented teeth. Regardless of the device used, a statistically significant difference was found between the measurements at 0.0 mm and -1.0 mm ($p < 0.05$).

Table 2: File tip position relative to the apical foramen for measurements performed to 0.0

Distance from apical foramen (mm)	Root ZX II		Propex II + MPAS	
	n	%	n	%
< -0.51*	0	0.0	0	0.0
-0.5 to -0.01*	3	50.0	5	83.3
0.00	1	16.7	1	16.7
0.01 to 0.5	2	33.3	0	0.0
> 0.51	0	0.0	0	0.0

Table 3: File tip position during measurements performed short of the apical foramen (-1.0 mm)

Distance from apical foramen (mm)	Root ZX II		Propex II + MPAS	
	n	%	n	%
< -2.01*	0	0.0	0	0.0
-2.0 to -1.51*	0	0.0	0	0.0
-1.5 to -1.01*	0	0.0	1	16.7
-1.00	0	0.0	0	0.0
-0.99 to -0.50	0	0.0	3	50.0
-0.49 to 0.0	5	83.3	2	33.3
> 0.01	1	16.7	0	0.0

DISCUSSION

The present study aimed to verify *ex vivo* the ability of devices that combine EFLs for determining the root canal length with electric motors for mechanical instrumentation of the root canal system, allowing precise control of the apical extent during the root canal preparation. Although care should be taken before extrapolating the results of *ex vivo* studies to clinical reality, studies have shown that the methodology used in this study was able to faithfully reproduce what was found in clinical conditions [23].

The methodological procedures performed in the present study do not differ from previous researches [4 6 7 1417]. However, special attention was given to standardization of the specimens and the used preparation levels. Relative to the standardization, the main aspect was AF calibration. It was performed to achieve similar apical adjustment for the employed rotary files; this adjustment was appointed as an important tool for increasing the EFL precision rates [5 6 9]. As for the apical preparation level (i.e., the working length {WL}), although the instrumentation is commonly established at 1.0 mm before the AF [4 7 21 22], recent studies presented a decrease of precision rates of ELFs at this level [7 9 17]. Considering this information, two WLs were employed at the present study: 1.0 mm before the AF, and at the AF, a reference point at which EFLs present their highest precision [7 9 17]; in both cases were used the same Protaper F3 file.

Until the present study was conducted, no study evaluated the accuracy of Propex II in monitoring the working length during mechanical preparation at any level. The MPAS/10R combination allowed the use of this apparatus throughout the mechanized preparation of the canal system. The results obtained by this association were compared to those by Root ZX II.

In the present study, Root ZX II presented extremely accurate results when controlling the extent of root canal preparation, when set to reach the AF. In all specimens of this group, the device pointed apical extension of instrumentation within the employed margin of tolerance (±0.5 mm). These results disagree with some previous studies that used its AAR control, presenting values around 50% or lower[3 18]. This disagreement may be related to the employed apical limit (the AF instead of 0.5 mm before the AF). Insertion until reaching the AF may have favored a better interpretation of the resistive factor when processing the data during the mechanical preparation, as demonstrated by Vasconcelos et al. [7]. When evaluating the results obtained at 1.0 mm short of the AF (-1.0 mm), theoretically defining the area of the pulp stump and apical stop, Root ZX II erroneously determined the canal length in all specimens, presenting a significant difference when compared to other groups.

Propex II, when used with the MPAS-10R showed 100% accuracy when set to reach the AF, considering the margin of error. However, when the working length was limited to -1.0 mm, it presented a decrease in accuracy, offering mean errors statistically different from those provided by this device at the AF; in every case better than those provided by Root ZX II at the same limit. This could be related to differences between their operating methods. Such results are difficult to compare with the published literature since this equipment has not been previously employed with a monitoring function. Considering the monitoring function as previously described, Propex II presented results slightly lower than those found in its conventional application [2] [15] [17]. This finding may be related to the existence of some delay between the EFL determination and the interruption of file penetration into the root canal.

It is important to emphasize the accuracy of the tested hybrid devices. Root ZX II, in AAR function and Propex II/MPAS-10R combination, both enabled monitoring and provided excellent reliability as tools for working length control during mechanical root canal preparation, as long as calibrated to the AF (0.0 mm). However, the results show great difficulty in maintaining accuracy when restricted to 1.0 mm short from the AF. Such findings highlight the reliability of these devices to control and/or monitor the extent of the root canal preparation; yet, they lose accuracy in positions below the AF. Their determinations in these cases should be confirmed, improving accuracy that may improve success rates of endodontic treatments.

Thus, considering the conditions of this study, the tested devices/ combinations were extremely accurate in maintaining the apical extent of mechanical instrumentation when used up to 0.0 mm from the AF. When limited to -1.0 mm from the AF, the devices presented reduced accuracy; with Root ZX II AAR system failing to indicate the correct limit in all cases.

ACKNOWLEDGEMENTS

This work was supported by FUNCAP (#PJP-0072-00098.01.00/12). The authors deny any conflicts of interest related to this study.

REFERENCES

1. Felippe WT, Felippe MCS, Reyes Carmona J, Crozoe FCI, Alvisi BB.*Ex vivo*1. evaluation of the ability of the Root ZX II to locate the apical foramen and to control the apical extent of rotary canal instrumentation. Int Endod J 2008;41:502-507.

2. Siu C, Marshall JG, Baumgartner JC. An *in vivo*2. comparison of the Root ZX II, the Apex NRG XFR, and Mini Apex Locator by using rotary nickel-titanium files. J Endod 2009;35:962-965.

3. Fadel G, Piasecki L, Westphalen VPD, Silva Neto UX, Fariniuk LF, Carneiro E. An *in vivo*3. evaluation of the Auto Apical Reverse function of the Root ZX II. Int Endod J 2012;45:950-954.

4.   Bernardes RA, Duarte MAH, Vasconcelos BC, Moraes IG, Bernardineli N, Garcia RB, et al.. Evaluation of precision length determination with 3 electronic apex locators: Root ZX, Elements Diagnostic Unit, RomiApex D-30. Oral Surg Oral Med Oral Pathol Oral Radiol Endod 2007;104:e91-e94.

5.   Camargo EJ, Zapata RO, Medeiros PL, Bramante CM, Bernardineli N, Garcia RB, et al.. Influence of preflaring on the accuracy of length determination with four electronic apex locators. J Endod 2009;35:1300-1302.

6.   Vasconcelos BC, Matos LA, Pinheiro-Júnior EC, Menezes AST, Vivacqua-Gomes N. Ex vivo6. accuracy of three electronic apex locators using different apical file sizes. Braz Dent J 2012;23:199-204.

7.   Vasconcelos BC, Frota LMA, Souza TA, Bernardes RA, Duarte MAH. Evaluation of the maintenance of the apical limit during instrumentation with hybrid equipment in rotary and reciprocating modes. J Endod 2015;41:682-685.

8.   Beltrame AP, Triches TC, Sartori N, Bolan M. Electronic determination of root canal working length in primary molar teeth: an in vivo8. and ex vivo8. study. Int Endod J 2011;44:402-406.

9.   Stober EK, Duran-Sindreu F, Mercandé M, Vera J, Bueno R, Roig M. An evaluation of Root ZX and iPex apex locators: An in vivo9. study. J Endod 2011;37:608-610.

10. Paludo L, Souza SL, Só MV, Rosa RA, Vier-Pelisser FV, Duarte MA. An in vivo10. radiographic evaluation of the accuracy of Apex and iPex electronic apex locators. Braz Dent J 2012;23:54-58.

11. Pereira KF, Silva PG, Vicente FS, Arashiro FN, Coldebella CR, Ramos CA. An in vivo11. study of working length determination with a new apex locator. Braz Dent J 2014;25:17-21.

12. Vasconcelos BC, Araujo RBR, Silva FCFA, Luna-Cruz SM, Duarte MAH, Fernandes CAO. In vivo12. accuracy of two electronic apex locators based on different operation systems. Braz Dent J 2014;25:12-16.

13. Nekoofar MH, Ghandi MM, Hayes SJ, Dummer PMH. The fundamental operating principles of electronic root canal length measurement devices. Int Endod J 2006;39:595-609.

14. D'Assunção FL, Albuquerque DS, Salazar-Silva JR, Santos VC, Sousa JC. Ex vivo14. evaluation of the accuracy and coefficient of repeatability of three electronic apex locators using a simple mounting model: a preliminary report. Int Endod J 2010;43:269-274.

15. Mancini M, Felici R, Conti G, Constantine M, Cianconi L. Accuracy of three electronic apex locators in anterior and posterior teeth: an *ex vivo*15. study. J Endod 2011;37:684-687.

16. Jung IY, Yoon BH, Lee SJ. Comparison of the reliability of "0.5" and "APEX" mark measurements in two frequency-based electronic apex locators. J Endod 2011;37:49-52.

17. Vasconcelos BC, Bueno MM, Luna-Cruz SM, Duarte MAH, Fernandes CAO. Accuracy of five electronic foramen locators with different operating systems: an *ex vivo*17. study. J Appl Oral Sci 2013;21:132-137.

18. Jacobson SJ, Westpalhen VPD, da Silva-Neto UX, Fariniuk LF, Picoli F, Carneiro E. The accuracy in the control of the apical extent of rotary canal instrumentation using Root ZX and Protaper instruments: an *in vivo*18. study. J Endod 2008;34:1342-1345.

19. Kuttler Y. Microscopic investigation of root apexes. J Am Dent Assoc 1955;50:544-552.

20. Riccuci D, Langeland K. Apical limit of root canal instrumentation and obturation, part 2. Histological study. Int Endod J 1998;31:394-409.

21. Souza RA. The importance of apical patency and cleaning of the apical foramen on root canal preparation. Braz Dent J 2006;17:6-9.

22. Gonzalez-Sanchez JA, Duran-Sidreu F, Noe S, Mercande M, Boig M. Centring ability and apical transportation after overinstrumentation with Protaper Universal and Profile Vortex instruments. Int Endod J 2012;45:542-551.

23. Duran-Cidreu F, Stober E, Mercandé M, Vera J, Garcia M, Bueno R, et al.. Comparison of *in vivo*23. and *in vitro*23. readings when testing the accuracy of the Root ZX apex locator. J Endod 2012;38:236-239.

Chapter 4

THE EFFECT OF INSTRUMENTAL TIMBRE ON INTERVAL DISCRIMINATION

Jean Mary Zarate , Caroline R. Ritson , David Poeppel

Department of Psychology, New York University, New York, New York, United States of America

ABSTRACT

We tested non-musicians and musicians in an auditory psychophysical experiment to assess the effects of timbre manipulation on pitch-interval discrimination. Both groups were asked to indicate the larger of two presented intervals, comprised of four sequentially presented pitches; the second or fourth stimulus within a trial was either a sinusoidal (or "pure"), flute, piano, or synthetic voice tone, while the remaining three stimuli were all pure tones. The interval-discrimination tasks were administered parametrically to assess performance across varying pitch distances between intervals ("interval-differences"). Irrespective of timbre, musicians displayed a steady improvement across interval-differences, while non-musicians only demonstrated enhanced interval discrimination at an interval-difference of 100 cents (one semitone in Western music). Surprisingly, the best discrimination performance across both groups was observed with pure-tone intervals, followed by intervals containing a piano tone. More specifically, we observed that: 1) timbre changes within a trial affect interval discrimination; and 2) the broad spectral characteristics of an instrumental timbre may influence perceived pitch or interval magnitude and make interval discrimination more difficult.

INTRODUCTION

The ability to perceive changing pitch in sounds is crucial for both speech and music. The contour of pitch changes in speech can determine the linguistic-communicative intent of a sentence (e.g., interrogative versus declarative versus imperative) and its affective content (happy, angry, sad, etc.), or—at the word level—distinguish lexical-semantic meanings in tonal languages [1].

The pitch contour in music outlines the melody. In more detail, the melodic contour can be subdivided into particular frequency ratios or intervals that have specific labels (and functions) in Western music composition, such as the minor third, perfect fifth, or the major seventh.

Since pitch intervals serve such a fundamental role in music, numerous studies have investigated the ability to discriminate pitch intervals in Western musical contexts (for a comprehensive review, see [2]). These experiments included tasks such as interval categorization or discrimination of interval magnitudes at or around musically relevant intervals[3], [4], [5], [6], [7], correcting mistuned intervals [8], [9], and assessment of performance intonation [10], [11]. It is likely that the explicitly musical contexts of these experiments—in which the discrimination tasks were based on musically relevant intervals—may have given musicians a significant advantage over non-musicians. In a recent experiment [12], we reduced the musical context by choosing frequencies that were not easily assigned to note names and interval magnitudes that are not often used in Western music (e.g., 25, 50, 75 cents), except one interval at 100 cents (a semitone). People with extensive musical expertise exhibited interval-discrimination thresholds of 100 cents, and non-musicians displayed larger thresholds[12], which: 1) corroborates McDermott et al.'s findings obtained with an adaptive procedure[13], and 2) suggests that an explicitly musical context in these studies may not influence the basic interval-discrimination thresholds in these groups. These thresholds may be established via repeated exposure to similar intervals in Western music and languages [14].

While musically relevant frequencies or interval magnitudes may not affect interval-discrimination thresholds, changes in a tone's frequency spectrum—which creates a particular timbre—may influence pitch and/or interval perception (see Figure 1) [15], [16]; however, earlier experiments have yielded conflicting accounts of these effects in musicians and non-musicians. When music students were asked to tune musical intervals containing pure or synthetic complex tones, Rakowski determined that regardless of tone timbre, melodic intervals of a minor third or less are judged as even smaller in size than their actual pitch magnitude, and conversely larger intervals are perceived as bigger that their magnitude [17]. In contrast, Russo and Thompson found that timbre affected the perceived size of a melodic interval for both musicians and non-musicians, depending on whether synthetic timbre changed from a dull to a brighter sound (or the reverse manipulation) between the two tones [18]. Spiegel and Watson[19] and Micheyl et al. [20] reported that both musicians and non-musicians had better two-tone discrimination thresholds with synthetic complex tones than with pure tones; they argued that the enhanced frequency discrimination observed with complex tones, which are closer to real instrumental timbres that musicians hear during training, may be generalized to artificial pure tones. Demany and Semal suggested that this generalization is only partial, since they found that pitch discrimination abilities may be specific for the timbres used during training [21]. Finally, McDermott and his colleagues found that musicians and non-musicians had similar pitch- and interval-discrimination thresholds for both synthetic complex and pure-tone stimuli[13].

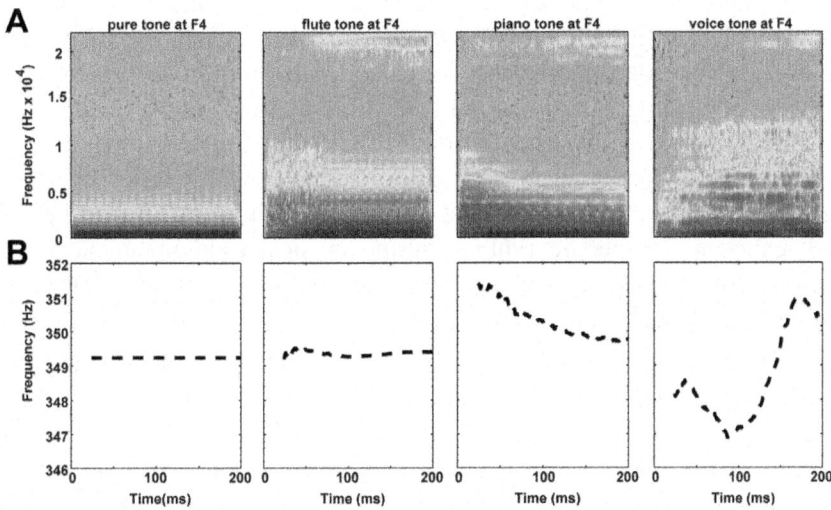

Figure 1. Spectrograms (A) and fundamental frequency traces (B) of the four tones at F4 (target base frequency of 349.23 Hz).

The variations in spectral energy influence the estimated fundamental frequency for each of the instrumental tones.

doi:10.1371/journal.pone.0075410.g001

Given these disparate accounts of timbral effects on interval discrimination, we designed this study to assess the phenomenon employing different levels of controls. Most auditory discrimination studies have typically used pure or synthetic complex tones; here, we employed more naturalistic, instrument sound samples to assess the effects of timbre. Non-musicians and musicians were asked to indicate the larger of two presented intervals. We manipulated the timbre in one of the four presented stimuli (either pure, flute, piano, or synthetic voice) per trial; the other three stimuli were pure tones, as in a previous experiment [12]. We altered only one out of four tones to determine whether the introduction of a different timbre would alter the perceived pitch of a tone and, therefore, the perceived interval size.

Additionally, we only changed one note per trial to prevent any confounding variables, such as interactions between two different non-sinusoidal timbres (one in each interval) and perceived pitch and/or interval size. Unlike our previous experiment, we sought to make this task more musically relevant by employing a base frequency (349.23 Hz or F4) and a large set of intervals that could be assigned to Western conventions: 100, 200, 300, 400, 500, 600 cents (i.e., the minor and major seconds, minor and major thirds, fourth, and the tritone). Based on earlier research [12], [13],[19], [20], [22], we predicted that the musicians would perform better on this discrimination task, due to their training-enhanced auditory skills. We also hypothesized—based on Spiegel and Watson's and Micheyl et al.'s reports of improved discrimination with complex tones—that instrumental timbres may improve all subjects' performances, since instrumental sounds are more naturalistic than pure tones that are only encountered in a laboratory setting. Finally, based on Demany and Semal's (2002) suggestion that enhanced auditory skills may be linked with exposure to specific timbres during musical training, musicians may exhibit greater improvement in interval discrimination during instrumental-timbre trials than non-musicians.

To summarize, we designed the present experiment to determine whether employing more naturalistic, instrumental timbres would improve interval discrimination in non-musicians and musicians, relative to only pure-tone stimuli. As detailed below, musicians discriminated between intervals better than non-musicians across all timbres, and interval discrimination was best with pure tones.

METHODS

Ethics Statement

All testing was performed with the subjects' informed written consent and in accordance with procedures approved by the NYU University Committee on Activities Involving Human Subjects.

Subjects

A total of 29 subjects were recruited from the New York University (NYU) community and surrounding areas. All subjects (mean age =24.8 years, SD=6.56 years) were right-handed and had normal hearing. All subjects were categorized as non-musicians or musicians according to self-report of musical experience, as assessed by an in-house survey. Fourteen non-musicians (7 female) had minimal musical experience (mean =0.78 years, SD=0.66 years) and did not play music regularly at the time of study. Fifteen musicians (7 female) had an average of 11.7 years of musical experience (SD=5.83 years) and were practicing or performing music at the time of study. None of the subjects reported having absolute pitch.

Stimuli

We used MATLAB (Mathworks, Natick, MA, USA) to create sinusoidal tones at a base frequency of 349.23 Hz, which corresponds to an F4 in Western music. The instrument sounds (piano, flute, voice) were MIDI-generated (Musical Instrument Digital Interface) from a Yamaha YPT-220 keyboard (Yamaha Corporation of America, Buena Park, CA, USA) at this same base frequency. All instrument sounds included the attack (or onset) of the sound and had no vibrato. Additional sinusoidal and instrumental tones were generated at specific pitch distances—50 to 600 cents at 25-cent increments—from this base frequency. All tones (200-ms duration, 16-bit depth, 44100-Hz sampling frequency) were gated with 7-ms cosine ramps in MATLAB and then normalized to 0.8 dB in Audacity (open-source freeware,http://audacity. sourceforge.net/). Figure 1 displays the spectrograms and the fundamental frequency (F0) estimated with YIN [23] for each timbre at F4. Compared to the pure tone, all instrumental timbres have more energy across a broader span of frequencies (including harmonics of the base frequency). Additionally, the flute and synthetic voice tones have more diffuse onsets, relative to the sharper onsets of the piano and pure tones. Finally, the frequency range of spectral energy increases from the piano to the flute tones, with the synthetic voice timbre displaying the broadest frequency distribution of energy compared to all other tones. The various changes in spectral energy apparently result in slight fluctuations of estimated F0 in the piano and synthetic voice timbres (Figure 1B).

In MATLAB, tones were paired with a 50-ms gap of silence in between to create interval sizes ranging from 50 to 600 cents (respectively, a quarter-tone to a tritone in Western music), and intervals were combined (ISI=0.8, 0.9, or 1 s) to create individual test trials. We parametrically varied the magnitude differences between intervals within a trial from 0 to 100 cents (a semitone in Western music) in 25-cent increments. We chose 100 cents as the maximum interval-difference due to both its musical relevance and the observation that musicians' performances approach a ceiling of maximum accuracy at around this magnitude [12]. Zero-cent differences between intervals in a trial were included to observe whether subjects—when forced to guess— had a response bias based on timbre type. For all trials, the first tone of each interval was a pure tone at the base frequency of 349.23 Hz. In trials with an instrumental timbre, the instrument sound could occur either as the second or the fourth tone.

Experimental procedure

In a sound-attenuated booth, subjects sat in front of a lab computer and wore headphones (Sennheiser HD 380 Professional, Sennheiser Electronic Corporation, Wedemark, Germany), through which all auditory stimuli were delivered via MATLAB at a comfortable level (~77.5 dB SPL). A 10-trial demonstration was presented at the beginning of the session to familiarize subjects with the different sounds presented during the experiment. Prior to interval discrimination, pitch-discrimination thresholds were determined with pure tones at 349.23 Hz in a "2 down – 1 up" staircase procedure [24], implemented as part of the MLP toolbox for auditory psychophysical testing [25]. After discrimination-threshold testing, subjects were presented with intervals in a two-alternative forced-choice design, indicating by button press which pair contained the larger interval. Subjects received visual feedback ("correct" or "incorrect") on the computer screen after making each decision. There were at least 15 trials of each interval type (pure, flute, piano, and voice) for each interval-difference, all presented in a pseudo-randomized order. In total, there were 5 blocks with 100 trials each, and subjects were allowed to take a short break between blocks.

Analyses

Subjects' performances were measured as percent-correct scores for each of the interval-differences (25–100 cents). We also calculated d-prime and βnormalized values to measure detector sensitivity and response bias, respectively [26], [27], [28]. The hit and false alarm (FA) rates, d-prime (d'), and βnormalized values were calculated as follows:

1. Hit =H(# times 1st pair was chosen/# trials with larger 1st pair) = score for 1st-pair trials.

2. FA =1 – H(# times 2nd pair was chosen/# trials with larger 2nd pair) =1 – score for 2nd-pair trials.

3. $d' = Zscore(Hit) - Zscore(FA)$

4. $\beta = -0.5 * (Zscore(Hit) + Zscore(FA))$

5. $\beta_{normalized} = \beta/d'$; 0= no bias; negative values = bias towards selecting 1st pair; positive values = bias towards selecting 2nd pair.

For equal-interval trials (0-cent difference between intervals), response bias was calculated as a proportion of the total number of trials in which subjects selected the interval with an instrumental timbre, instead of the pure-tone interval; higher proportions reflect a stronger bias towards selecting the instrumental-timbre intervals.

Pitch-discrimination thresholds (in cents), percent-correct scores, d' values, βnormalized values, mean reaction time, and standard deviation of reaction times were analyzed using an independent-samples t-test or repeated-measures analyses of variance (ANOVA). Tukey's Honestly Significant Difference test and planned comparisons were used for post-hoc analyses of significant main effects and interactions, respectively.

RESULTS

Effects of musical expertise on pitch-discrimination thresholds

An independent samples t-test performed on the musicians' and non-musicians' pitch-discrimination thresholds measured at F4 (349.23 Hz) determined that musicians' thresholds (mean ± SEM =14.6±2.7 cents) were significantly lower than non-musicians' thresholds (44.1±9.7 cents) as expected [t(27) =1.22, p<0.05].

Effects of musical expertise and timbre on interval-discrimination accuracy

Figure 2 depicts the significant results of a three-way repeated-measures ANOVA performed on percent-correct scores, with group as the between-subject factor and timbre and interval-difference as the repeated within-subject factors. The analysis revealed significant main effects of group [F(1,27) =33.64, p<0.001], timbre [F(3,81) =22.87, p<0.001], and interval-difference [F(3,81) =88.47, p<0.001], and significant two-way interactions between group and timbre [F(3,81) =5.75, p<0.01], group and interval-difference [F(3,81) =4.65, p<0.01], and timbre and interval-difference [F(9,243) =8.44, p<0.001]. No other interactions were significant.

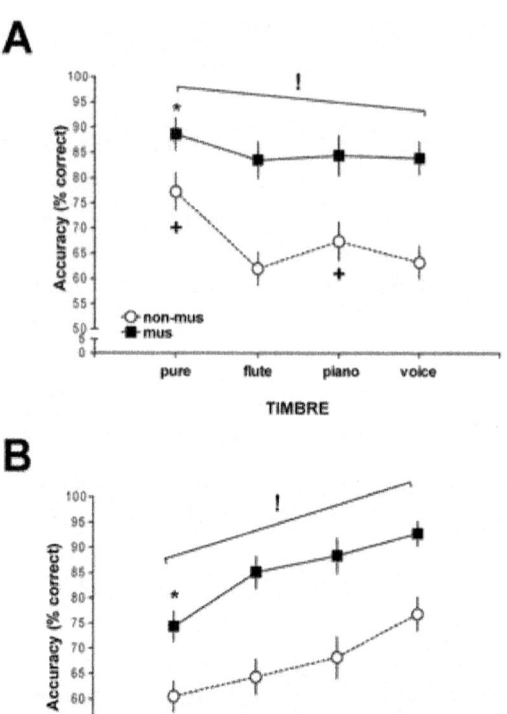

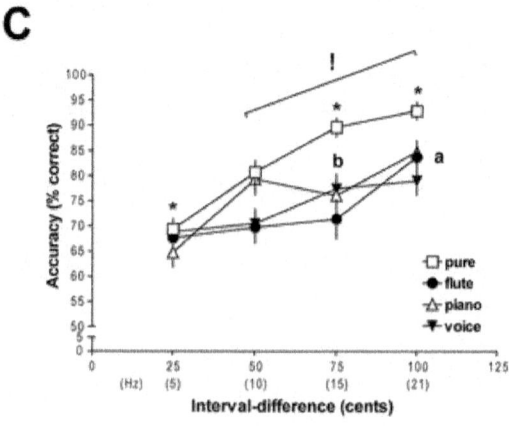

Figure 2. Interval-discrimination accuracy.

(A) Mean ± SEM percent-correct scores of musicians and non-musicians for interval discrimination, across the four timbre types and collapsed across all interval-differences. Musicians performed more accurately than non-musicians across all timbres (indicated by!, p<0.001), but were marginally most accurate with pure-tone intervals (denoted by *,p<0.06). Non-musicians showed more accurate interval discrimination with pure and piano tones than with the other two timbres (marked by +, p<0.05). (B) Mean ± SEM percent-correct scores of musicians and non-musicians for each interval-difference collapsed across timbre types. Musicians were more accurate than non-musicians overall (marked by!, p<0.001). Both musicians (denoted by *, p<0.001) and non-musicians (indicated by +, p<0.05) displayed the least accuracy at the 25-cent interval-difference. (C) Mean ± SEM percent-correct scores for each timbre type at each interval-difference, averaged across both groups. All subjects discriminated pure-tone intervals more accurately than flute- and voice-tone intervals at all interval-differences except 25 cents (shown by!, p<0.001), and better than piano-tone intervals at all interval-differences except 50 cents (marked with *, p<0.06). Flute-tone discrimination only improved at 100 cents (shown by a, p<0.001), while voice-tone performance significantly improved at interval-differences of 75 cents and higher (indicated by b,p<0.01).

doi:10.1371/journal.pone.0075410.g002

Planned comparisons performed on the group-by-timbre interaction determined that musicians discriminated intervals of all timbres more accurately than non-musicians (Figure 2A;ps<0.001), which reiterated the significant group main effect, but musicians discriminated intervals with all pure tones marginally better than with other timbres (ps<0.06). Non-musicians on the whole performed more accurately with pure and piano tones, which may be due to their sharper onsets and/or relatively compact distribution of sound energy (see Figure 1), compared to flute and synthetic voice timbres (ps<0.05). Planned comparisons on the two-way interaction between group and interval-difference showed that musicians discriminated between intervals more accurately at all interval-differences than non-musicians (Figure 2B; ps<0.001). Both groups showed a significant, steady improvement in accuracy as interval-differences increased, and had the worst discrimination accuracy at the 25-cent interval-difference compared to all other magnitudes (ps<0.001 for musicians; ps<0.05 for non-musicians).

Planned comparisons on the timbre-by-interval-difference interaction determined that discrimination with pure-tone intervals significantly improved at each larger interval-difference (Figure 2C; ps<0.05). Pure-tone discrimination was also better than with flute- and voice-tone intervals at all interval-differences except 25 cents (ps<0.001), and marginally more accurate than with piano-tone intervals at all interval-differences other than 50 cents (ps<0.06). This suggests that in general, interval discrimination with pure tones was better than with any other timbre. Performance with flute tones only significantly improved at the 100-cent interval-difference (ps<0.001). Discrimination accuracy with piano-tone intervals improved as interval-differences increased (ps<0.05), but with no significant change in accuracy between interval-differences of 50 and 75 cents. Voice-tone interval discrimination improved significantly beginning at the 75-cent interval-difference (ps<0.01); there were no significant changes in accuracy between 25- and 50-cent interval-differences and between 75- and 100-cent interval-differences. Notably, voice-tone interval discrimination was worse at 100 cents than with any other timbre (ps<0.01).

Effects of musical expertise and timbre on interval-discrimination sensitivity

Figure 3A shows the results of a three-way repeated-measures ANOVA (group by timbre by interval-difference) performed on d' prime values to assess the influence of musical expertise on interval-discrimination sensitivity. The analysis resulted in significant main effects of group [F(1,27) =23.28, p<0.001], timbre [F(3,81) =12.23, p<0.001], and interval-difference [F(3,81) =28.04, p<0.001], and significant two-way interactions between group and interval-difference [F(3,81) =7.02, p<0.001] and timbre and interval-difference [F(9,243) =2.47, p<0.05].

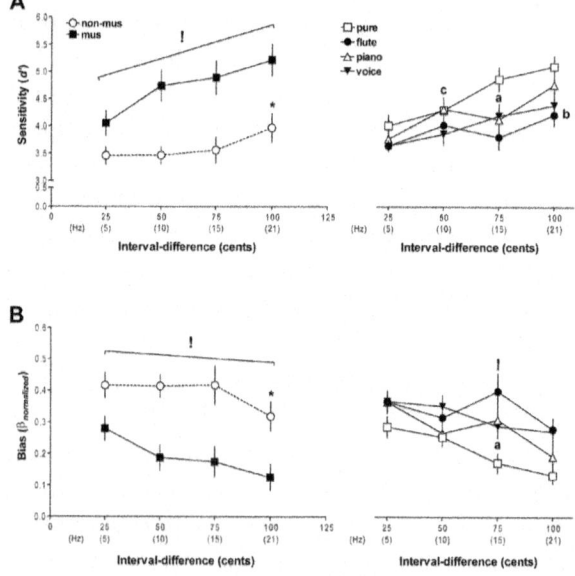

Figure 3. Interval-discrimination sensitivity and response bias.

Planned comparisons on the group-by-interval-difference interaction determined that musicians displayed more discrimination sensitivity than non-musicians across all interval-differences (Figure 3A, left; ps<0.01), as reflected by the main group effect. Among musicians, sensitivity improved as a function of increasing interval-difference (ps<0.01) irrespective of timbre, with the exception of no significant change in sensitivity between interval-differences of 50 and 75 cents. Within the non-musician group, discrimination sensitivity was only significantly enhanced at an interval-difference of 100 cents (ps<0.01); there were no other changes in sensitivity seen among smaller interval-differences (ps>0.5).

(A) Left: Mean ± SEM d' values of musicians and non-musicians across all timbres at each interval-difference. Musicians exhibited more discrimination sensitivity across all interval-differences, compared to non-musicians (denoted by!, p<0.01). Across interval-differences, on-musicians only showed enhanced discrimination sensitivity at a difference of 100 cents (marked by *, p<0.01). Right: Mean ± SEM d' values for each timbre at each interval-difference, averaged across both groups. Listeners exhibited enhanced discrimination sensitivity at interval-differences of 75 cents with pure and voice tones (indicated with a, ps<0.05), 100 cents with flute tones (marked with b, p<0.001), and 50 cents with a piano timbre (denoted by c, p<0.06). (B) Left: Mean ± SEMβ$_{normalized}$ values (measure of response bias) for musicians and non-musicians at each interval-difference, collapsed across all timbre types. In general, non-musicians showed greater bias towards selecting the second interval as the larger interval than non-musicians (shown with!, p<0.01), and this bias only significantly decreased within the group at the 100-cent interval-difference (indicated by *, p<0.001). Right: Mean ± SEMβ$_{normalized}$ values for each timbre at all interval-differences, averaged across both groups. Response bias decreased significantly with pure and voice-tones at 75 cents (marked by a, ps<0.05). Interestingly, response bias was highest at a 75-cent interval-difference with flute tones, compared to all other timbres (shown by!, p<0.05).

doi:10.1371/journal.pone.0075410.g003

Planned comparisons on the interaction between timbre and interval-difference revealed that d'values increased significantly across both groups for both pure- and voice-tone interval discrimination at interval-differences of 75 cents and larger (Figure 3A, right; ps<0.05). The d'values were higher for pure-tone interval discrimination than for flute-tone discrimination at all interval-differences (ps<0.05) except 50 cents, piano-tone discrimination at only 75 cents (p<0.001), and discrimination with voice tones overall (ps<0.05). Among the instrumental timbres, sensitivity for piano tones was marginally higher at 50–100 cents than for flute tones (ps<0.08) and at 50- and 100-cent interval-differences than for voice tones (ps<0.07). Sensitivity to flute-tone intervals increased significantly between 25 and 50 cents (p<0.01) and at the 100-cent interval-difference (ps<0.001). The d' values for piano-tone discrimination marginally increased at interval-differences of 50 cents and higher (ps<0.06), with no significant changes in sensitivity between 50- and 75-cent interval-differences.

Overall, interval-discrimination performance was best with pure-tone intervals, except at the smallest interval-differences of 25 cents; changes in timbral qualities did not enhance interval discrimination at this very small magnitude. Discrimination sensitivity was poorest with flute tones, whereas the poorest accuracy was observed with voice-tone intervals—accuracy with voice tones was still significantly lower than all other timbres at the 100-cent interval-difference. The discrepancy between accuracy and discrimination sensitivity may be explained by response bias, as discussed below.

Effects of musical expertise and timbre on response bias during interval discrimination

Figure 3B shows results from a three-way repeated measures ANOVA (group by timbre by interval-difference) on $\beta_{normalized}$ values to determine the effects of musical training and timbre on response bias. We found significant main effects of group [$F(1,27) = 21.70$, $p<0.001$], timbre [$F(3,81) = 16.70$, $p<0.001$], and interval-difference [$F(3,81) = 27.89$, $p<0.001$], as well as significant group-by-interval-difference [$F(3,81) = 5.61$, $p<0.01$] and timbre-by-interval-difference interactions [$F(9, 243) = 3.57$, $p<0.001$].

Planned comparisons on the group-by-interval-difference interaction revealed that non-musicians showed greater bias than musicians towards choosing the second interval as the larger interval across all interval-differences (Figure 3B, left; ps<0.01), as also indicated by the significant group main effect. Musicians' response bias decreased as interval-differences grew (ps<0.05), except for no change in response bias between interval-differences of 50 and 75 cents. Among non-musicians, response bias only significantly decreased at the 100-cent interval-difference (ps<0.001); no other significant changes in bias were seen at smaller magnitudes (ps>0.8). This result is mirrored by non-musicians' significant increase in discrimination sensitivity at only a 100-cent interval-difference (see results for d' values above).

Planned comparisons performed on the timbre-by-interval-difference interaction determined that response bias significantly decreased with pure tones as interval-differences increased to 75 cents and larger (Figure 3B, right; ps<0.05). There was also less response bias with pure-tone intervals compared to flute and voice tones at all interval-differences (ps<0.05), and relative to piano tones at 25- and 75-cent interval-differences (ps<0.01). During interval discrimination with flute tones, response bias was higher at interval-differences of 25 and 75 cents than at 50 and 100 cents (ps<0.05); notably, bias unexpectedly increased between interval-differences of 50 and 75 cents (p<0.05). The bias observed at 75 cents with flute tones was the largest compared to all other timbres (ps<0.05), which may explain why sensitivity was worst with flute tones and only improved at a 100-cent interval-difference. Response bias with piano-tone intervals reduced as a function of increasing interval-difference (ps<0.05), with the exception of no significant bias change between interval-differences of 50 and 75 cents. During interval discrimination with voice tones, response bias decreased significantly at 75 cents and larger (ps<0.05); bias did not change between 25- and 50-cent and between 75- and 100-cent differences between intervals.

To assess whether timbre influenced response bias with equal-magnitude intervals as subjects were forced to guess the larger interval, we performed an additional three-way repeated measures ANOVA on response-bias scores from the 0-cent interval-difference trials; response bias was measured as the proportion of trials in which subjects selected the interval with an instrumental timbre instead of the pure-tone interval. The analysis resulted in a significant group effect [$F(1,27) = 16.10$, p<0.001]—non-musicians exhibited a greater bias (mean ± SEM: 0.66±0.04) than musicians (0.53±0.03) towards selecting intervals with instrumental timbres when there was no difference in interval magnitude.

Effects of musical expertise and timbre on reaction time during interval discrimination

Figure 4 displays results from analyses on mean reaction times obtained during interval discrimination. We found significant main effects of timbre [$F(3,81) = 4.68$, p<0.01] and interval-difference [$F(3,81) = 23.77$, p<0.001], and a significant group-by-interval-difference interaction [$F(3,81) = 6.40$, p<0.001]. Post-hoc tests performed on the timbre main effect determined that all subjects responded marginally faster during interval discrimination with pure tones than any other timbre (Table 1; ps<0.06). Planned comparisons on the group-by-interval-difference interaction revealed that musicians were marginally faster than non-musicians in discriminating intervals that were 100 cents apart (Figure 4; p<0.09), but not at any other interval-difference. Within the musician group, reaction times reduced as interval-differences increased (ps<0.01), except for no significant change in reaction time between interval-differences of 50 and 75 cents. Non-musicians responded more quickly at interval-differences of 100 cents than at 25 and 50 cents (Figure 4; ps<0.01).

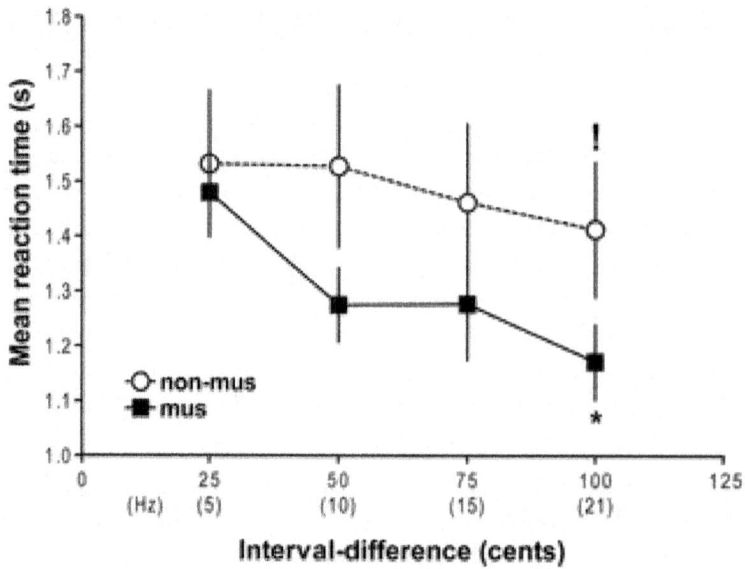

Figure 4. Mean ± SEM reaction times (in seconds) for musicians and non-musicians, averaged across all timbres.

Musicians answered marginally faster than non-musicians at a 100-cent interval-difference (marked by *, p<0.09). Within the non-musician group, reaction times were faster at an interval-difference of 100 cents, compared to 25 and 50 cents (indicated by!,p<0.01).

doi:10.1371/journal.pone.0075410.g004

Table 1. Mean ± SEM reaction times (in seconds) during interval discrimination with each timbre type.

Timbre	Mean Reaction Time ± SEM (s)
pure	1.329±0.066*
flute	1.428±0.089
piano	1.400±0.088
voice	1.399±0.078

Both musicians and non-musicians discriminated between pure-tone intervals faster than intervals with other timbres (marked by *, p<0.06), regardless of interval-difference.
doi:10.1371/journal.pone.0075410.t001

Analyses of the variability of reaction time (measured as a subject's standard deviation of reaction time) revealed a marginally significant main effect of timbre [$F(3,81) = 2.20$, $p < 0.1$]; post-hoc tests determined that response times were marginally more variable during flute-tone discrimination than during pure-tone discrimination (Table 2; $p < 0.08$); no other significant differences were found. Based on accuracy and sensitivity scores, interval discrimination with flute tones may be more difficult, and this may be accompanied by more variable response times.

Table 2: Average variability ± SEM of reaction time (in seconds) for each timbre type.

Timbre	Mean Variability of Reaction time ± SEM (s)
pure	0.540±0.060
flute	0.720±0.146 *
piano	0.669±0.177
voice	0.607±0.073

Across all interval-differences, all subjects showed more variable response times during interval discrimination with flute tones than with pure tones (indicated by *, $p < 0.08$).
doi:10.1371/journal.pone.0075410.t002

Analyses of the reaction times during the equal-interval trials resulted in a significant timbre effect [$F(3,81) = 4.06$, $p < 0.01$]—all subjects responded more quickly during flute-tone trials than during pure- and voice-tone trials (Table 3; $ps < 0.05$), but not compared to during piano-tone trials. Analyses of the variability of reaction time during equal-interval trials did not reveal any significant other main effects or interactions ($ps > 0.2$).

Table 3: Mean ± SEM of reaction time (in seconds) during equal-interval trials.

Timbre	Mean Reaction Time ± SEM (s)
pure	1.625±0.089
flute	1.469±0.068 *
piano	1.578±0.091
voice	1.600±0.088

Regardless of interval-difference, flute-tone intervals elicited shorter reaction times from all subjects, compared to intervals with pure and voice tones (shown with *, $p < 0.05$).
doi:10.1371/journal.pone.0075410.t003

DISCUSSION

As expected, musicians discriminated between intervals more accurately than non-musicians across all interval-differences. In general, accuracy (as measured by percent-correct scores) improved in both groups as interval-differences increased. When examining discrimination sensitivity (represented by d′ values), musicians displayed a significant increase in discrimination sensitivity across most interval-differences, while non-musicians exhibited an interval-discrimination sensitivity threshold of 100 cents. Originally, we hypothesized that instrumental sounds would enhance interval-discrimination performance in all subjects, relative to pure tones. However, although non-musicians displayed a stronger bias to select the interval with any instrumental sound among pairs of equal-magnitude intervals, both groups typically discriminated between intervals better and faster with pure tones than with any of the instrumental timbres; this evidence contradicts Micheyl and colleagues› report of enhanced auditory perception with complex timbres relative to pure tones [20]. Among the instrumental timbres, interval discrimination was best with piano tones (perhaps due to a sharper onset or attack), and intervals with flute tones (with diffuse tone onsets) elicited arguably the worst discrimination sensitivity.

Since both groups also took longer to respond during instrumental-timbre trials, it is possible that changing the timbre of only one out of four stimuli presented within a trial could have caused a distraction during discrimination; sequential tones with different timbres may be difficult to group together as intervals in our experiment [29]. Indeed, a previous study has suggested that the introduction of a new timbre can violate expectations that each successive sound will match the timbre of the previous one(s); in that study, this violation of expectation manifested as decreased discrimination accuracy [30]. Our observed interaction between timbre changes and decreased task performance is supported by Borchert et al.›s observations[31]: subjects had difficulty discriminating between two sequentially presented tones with different timbres, compared to two simultaneous tones with different timbres. Moreover, timbre seems to have a stronger interaction with pitch extraction/judgment and the evaluation of interval size when there is no tonal context (as in our interval-discrimination task), as opposed to tasks with a tonal reference point (or "key") [30]. In fact, pitch changes seem to be best perceived (regardless of timbre) when F0 changes by at least 4%; the perception of smaller F0 changes is more influenced by timbre changes [16]. In our task, two interval sizes and interval-differences have F0 changes of less than 4% of the base frequency (i.e., 25 and 50 cents from 349.23 Hz). Thus, our experimental design perhaps allowed for a stronger influence of timbral manipulation on pitch perception and interval discrimination than expected, which may have rendered interval comparisons with instrumental timbres more difficult. This may also explain the non-musicans› bias towards selecting the interval with an instrumental timbre as the larger interval, even when the interval magnitudes were equal; timbral changes may have altered non-musicians› perception of the instrumental tones› spectral centroid and consequently the interval magnitude (see [18]). In contrast, musicians are reported to be less susceptible to timbre-based illusions of interval size, making them less likely to perceive the instrumental interval as larger[18].

Although we expected an improvement in interval discrimination based on both instrumental timbre and musical expertise, such an interaction was not observed in this study. However, when comparing the present results to those of our previous study with only pure-tone intervals[12], the use of instrumental timbre with intervals based on a musically relevant frequency appears to improve interval discrimination in each group, specifically among our parametrically varied interval-differences. In our earlier study, we reported pure-tone interval-discrimination thresholds of 100 cents in musicians, and 150 cents in non-musicians, which echoed McDermott et al.›s (2010) findings with both pure- and complex-tone intervals. Surprisingly, the musicians enrolled in this study did not display a threshold in interval discrimination (when averaging across all timbres), but rather a steady, significant increase in performance as the interval-differences grew larger. In addition, the non-musicians here demonstrated a smaller interval-discrimination threshold of 100 cents (across all timbres), rather than the threshold seen in our earlier experiment with only pure tones. In general, these qualitative performance changes across the two studies suggest that timbre specifically from musical instruments may improve or aid interval discrimination. Whether this instrumental-timbre effect can be disentangled from the effect of implementing a musically relevant base frequency (F4), rather than the base frequencies employed in our earlier study or in McDermott et al.›s (2010) study, must be explored further in later research.

However, these interpretations of timbral effects should be taken cautiously, since the best interval discrimination was still observed with pure-tone stimuli, which may have been due partly to higher presentation rates of pure-tone intervals throughout the experiment, compared to instrumental-tone intervals; a practice effect with this particular timbre may have enhanced pure-tone interval discrimination. Nevertheless, compared to previous conflicting accounts of the timbral effects on interval discrimination, we observed that: 1) changes from one timbre to another, especially within the same trial, significantly affect interval discrimination, and 2) the varied spectral energy of instrumental timbre can alter pitch perception and/or interval discrimination.

ACKNOWLEDGMENTS

The authors thank Keith Doelling for his helpful feedback on this manuscript.

AUTHOR CONTRIBUTIONS

Conceived and designed the experiments: JMZ CRR DP. Performed the experiments: JMZ CRR. Analyzed the data: JMZ CRR. Wrote the paper: JMZ CRR DP.

REFERENCES

1.   Dowling WJ, Harwood D (1986) Music Cognition. New York: Academic Press. 1–258 .

2.   Burns EM (1999) Intervals, Scales, and Tuning. In: Deutsch D, The Psychology of Music. San Diego: Academic Press. 215–264.

3.   Burns EM, Ward WD (1978) Categorical perception–phenomenon or epiphenomenon: evidence from experiments in the perception of melodic musical intervals. J AcoustSoc Am 63: 456–468. doi: 10.1121/1.381737

4.   Hill TJ, Summers IR (2007) Discrimination of interval size in short tone sequences. J AcoustSoc Am 121: 2376–2383. doi: 10.1121/1.2697059

5.   Houtsma AJM (1968) Discrimination of frequency ratios. J AcoustSocAm 44: 383. doi: 10.1121/1.1970636

6.   Zatorre RJ (1983) Category-boundary effects and speeded sorting with a harmonic musical-interval continuum: evidence for dual processing. J Exp Psychol HumPerceptPerform 9: 739–752. doi: 10.1037/0096-1523.9.5.739

7.   Zatorre RJ, Halpern AR (1979) Identification, discrimination, and selective adaptation of simultaneous musical intervals. PerceptPsychophys 26: 384–395. doi: 10.3758/bf03204164

8.   Rakowski A (1976) Tuning of isolated musical intervals. JAcoustSocAm 59: S50. doi: 10.1121/1.2002737

9.   Ward WD (1954) Subjective musical pitch. J AcoustSocAm 26: 369–380. doi: 10.1121/1.1907344

10.   Dowling WJ (1978) Scale and contour: two components of a theory of memory for melodies. Psychol Rev 85: 341–354. doi: 10.1037/0033-295x.85.4.341

11.   Ward WD (1970) Musical perception. In: Tobias JV, Foundations of modern auditory theory. New York: Academic Press. 405–447.

12. Zarate JM, Ritson CR, Poeppel D (2012) Pitch-interval discrimination and musical expertise: is the semitone a perceptual boundary? J AcoustSoc Am 132: 984–993. doi: 10.1121/1.4733535

13. McDermott JH, Keebler MV, Micheyl C, Oxenham AJ (2010) Musical intervals and relative pitch: frequency resolution, not interval resolution, is special. J AcoustSoc Am 128: 1943–1951. doi: 10.1121/1.3478785

14. Han S, Sundararajan J, Bowling DL, Lake J, Purves D (2011) Co-variation of tonality in the music and speech of different cultures. PLoS One 6: e20160. doi: 10.1371/journal.pone.0020160

15. Moore BCJ, Glasberg BR (1990) Frequency discrimination of complex tones with overlapping and non-overlapping harmonics. The Journal of the Acoustical Society of America 87: 2163–2177. doi: 10.1121/1.399184

16. Singh PG, Hirsh IJ (1992) Influence of spectral locus and F0 changes on the pitch and timbre of complex tones. Journal of the Acoustical Society of America 92: 2650–2661. doi: 10.1121/1.404381

17. Rakowski A (1990) Intonation variants of musical intervals in isolation and in musical contexts. Psychology of Music 18: 60–72. doi: 10.1177/0305735690181005

18. Russo FA, Thompson WF (2005) An interval size illusion: the influence of timbre on the perceived size of melodic intervals. Percept Psychophys 67: 559–568. doi: 10.3758/bf03193514

19. Spiegel MF, Watson CS (1984) Performance on frequency-discrimination tasks by musicians and non-musicians. JAcoustSocAm 76: 1690–1695. doi: 10.1121/1.391605

20. Micheyl C, Delhommeau K, Perrot X, Oxenham AJ (2006) Influence of musical and psychoacoustical training on pitch discrimination. HearRes 219: 36–47. doi: 10.1016/j.heares.2006.05.004

21. Demany L, Semal C (2002) Learning to perceive pitch differences. J AcoustSoc Am 111: 1377–1388. doi: 10.1121/1.1445791

22. Kishon-Rabin L, Amir O, Vexler Y, Zaltz Y (2001) Pitch discrimination: are professional musicians better than non-musicians? J Basic ClinPhysiol Pharmacol 12: 125–143. doi: 10.1515/jbcpp.2001.12.2.125

23. de Cheveigné A, Kawahara H (2002) YIN, a fundamental frequency estimator for speech and music. J AcoustSoc Am 111: 1917–1930. doi: 10.1121/1.1458024

24. Levitt H (1971) Transformed up-down methods in psychoacoustics. J AcoustSoc Am 49: 467–477. doi: 10.1121/1.1912375

25. Grassi M, Soranzo A (2009) MLP: a MATLAB toolbox for rapid and reliable auditory threshold estimation. BehavRes Methods 41: 20–28. doi: 10.3758/brm.41.1.20

26. Dorfman DD, Alf E Jr (1968) Maximum likelihood estimation of parameters of signal detection theory–a direct solution. Psychometrika 33: 117–124. doi: 10.1007/bf02289677

27. Rosenblith WA, Stevens KN (1953) On the DL for frequency. JAcoustSocAm 25: 980–985.

28. Swets J (1982) Recent theoretical developments in signal-detection and recognition. Psychophysiology 19: 300.

29. Bregman AS (1990) Auditory scene analysis: The perceptual organization of sound. Cambridge: MIT Press.

30. Warrier CM, Zatorre RJ (2002) Influence of tonal context and timbral variation on perception of pitch. Percept Psychophys 64: 198–207. doi: 10.3758/bf03195786

31. Borchert EM, Micheyl C, Oxenham AJ (2011) Perceptual grouping affects pitch judgments across time and frequency. J Exp Psychol Hum Percept Perform 37: 257–269. doi: 10.1037/a0020670

Chapter 5

OPEN-SOURCE HARDWARE IS A LOW-COST ALTERNATIVE FOR SCIENTIFIC INSTRUMENTATION AND RESEARCH

Daniel K. Fisher[1], Peter J. Gould[2]

[1]USDA Agricultural Research Service, Stoneville, USA

[2]US Forest Service, Pacific Northwest Research Station, Olympia, USA

ABSTRACT

Scientific research requires the collection of data in order to study, monitor, analyze, describe, or understand a particular process or event. Data collection efforts are often a compromise: manual measurements can be time-consuming and labor-intensive, resulting in data being collected at a low frequency, while automating the data-collection process can reduce labor requirements and increase the frequency of measurements, but at the cost of added expense of electronic data-collecting instrumentation. Rapid advances in electronic technologies have resulted in a variety of new and inexpensive sensing, monitoring, and control capabilities which offer opportunities for implementation in agricultural and natural-resource research applications. An Open Source Hardware project called Arduino consists of a programmable microcontroller development platform, expansion capability through add-on boards, and a programming development environment for creating custom microcontroller software. All circuit-board and electronic component specifications, as well as the programming software, are open-source and freely available for anyone to use or modify. Inexpensive sensors and the Arduino development platform were used to develop several inexpensive, automated sensing and datalogging systems for use in agricultural and natural-resources related research projects. Systems were developed and implemented to monitor soil-moisture status of field crops for irrigation scheduling and crop-water use studies, to measure daily evaporation-pan water levels for quantifying evaporative demand, and to monitor environmental parameters under forested conditions. These studies demonstrate the usefulness of automated measurements, and offer guidance for other researchers in developing inexpensive sensing and monitoring systems to further their research.

INTRODUCTION

Scientific research requires the collection of data in order to study, monitor, analyze, describe, or understand a particular process or event. Data collection efforts are often a compromise, however, between the amount and type of measurements needed and the resources available to collect them. Manual measurements can be time-consuming and labor-intensive, resulting in data being collected at a low frequency, with long time intervals between measurements. If outdoor field research is involved, collection intervals can be irregular when labor is unavailable, on weekends or when other duties take priority for example, or when inclement weather does not permit visits to the field. Automating the data-collection process can reduce labor requirements and greatly increase the frequency and regularity of measurements, but at the cost of added expense of electronic data-collecting instrumentation.

A vast number of electronic solutions are available for automated sensing, monitoring, and collecting information, but several problems exist which can limit their application in research work and acceptance by research scientists. Features, capabilities, and prices of commercially available datalogging instrumentation can vary greatly, from inexpensive, low-resolution, limited-input devices to expensive, full-featured, multi-input instruments. Developed by private industry, monitoring equipment often contains proprietary technology that manufacturers do not wish to release, and is often designed to operate with only a particular manufacturer's sensors. The user can become locked into a particular manufacturer's systems or sensor technology due to high costs of the monitoring equipment making it cost-prohibitive to switch to a different vendor. If a range of different sensor information is desired, a single vendor may not supply all that is needed, and several monitoring systems may be required due to incompatible technologies. Since the scientific data-collection and monitoring market is small, private companies may be slow to innovate or introduce new technologies based solely on economic analyses. And to obtain sufficient quantities of data from an experiment, multiple sites and replicated treatments may be needed to satisfy observational and statistical requirements, which can quickly become cost-prohibitive.

Rapid advances in electronic technologies have resulted in a variety of new and inexpensive sensing, monitoring, and control capabilities. These rapidly evolving technologies provide researchers and practitioners with access to low-cost, solid-state sensors and programmable microcontroller-based circuits. Microcontrollers can be thought of as small, low-power, low-cost computers packaged within a single chip. The microcontroller runs a program that is created and uploaded by the user to operate different components within a circuit. The user can modify the program and change the function of the circuit without changing the circuit physically. Many types of sensors and auxiliary components, such as memory chips, clocks, and communications devices, are available which interface directly with microcontrollers, simplifying circuit designs and putting electronic design within reach of people with limited electronics background and knowledge. A number of microcontroller-based devices have been described in which the specific requirements of a research project dictated the development of customized monitoring systems with unique capabilities [1-5].

A further advancement in microcontroller-based sensing and monitoring relates not specifically to the design and development of the electronics and physical components, but to the idea of making the designs and development efforts freely available to all in order to facilitate and expand the adoption of the technologies. The rapid rise of the internet and accessibility of computer resources led to the concept of Open Source Software as a means to provide free and transparent access to computer code so that individuals could review, modify, improve, and distribute computer software (Open Source Initiative, http://www.opensource.org). In recent years, a similar effort was undertaken to enable the free and open sharing of hardware designs and projects so that, by sharing and collaborating with others who have similar interests and needs, innovation could occur more quickly, improvements could be suggested and incorporated, and more users could access the final product.

One such Open Source Hardware project resulted in the creation of a microcontroller-based development platform called Arduino [6]. The Arduino hardware consists of a programmable microcontroller mounted on a circuit board which provides convenient access to the microcontroller input/output pins and connectivity to a personal computer for programming and user interaction. The circuit board has a standardized size and physical configuration so that any Arduino-compatible boards can be interchanged. Standardized add-on boards (called shields) plug into the Arduino circuit board, and are used to expand the capabilities of the main board. The microcontroller is programmed via the Arduino Integrated Development Environment (IDE), in which the user creates the program instructions to operate the microcontroller and then downloads the program to the microcontroller. As an open-source hardware project, all circuitboard and electronic component specifications, as well as the IDE software, are freely available for anyone to use or modify. As a result, private manufacturers all around the world produce and offer inexpensive, standardized Arduino-compatible hardware with an extensive supply of features and capabilities. Researchers have begun to develop and implement devices based on the Arduino platform for a variety of applications [7-12], with ease of use, low cost, and standardized components and programming language cited as reasons for choosing the Arduino platform.

The objective of this paper is to introduce researchers and practitioners to potential applications of the opensource Arduino platform for implementation in research and monitoring applications. Specifically, we 1) describe the Arduino microcontroller development platform, 2) discuss examples of sensing and auxiliary circuit components available, and 3) demonstrate several datalogging devices developed for use in agricultural and natural-resources research.

COMPONENTS

Arduino Microcontroller Development Platform

The current standard Arduino development platform is based on an ATmega328 8-bit programmable microcontroller (Atmel Corporation, San Jose, CA USA). A printed-circuit board positions the microcontroller in a circuit so that the input/output (IO) pins are easily accessible. The microcontroller contains 32 kilobytes (KB) of flash memory for program storage and 1 KB of non-volatile data-storage memory. IO lines consist of 14 digital pins and 6 analog pins, which provide 6 channels of 10-bit analog-to-digital (A/D) conversion capability. The microcontroller contains many built-in features, including timer/counters, internal and external interrupts, serial and other communication-protocol capabilities, programmable watchdog timer, and low-power, energy-saving modes.

Versions of the Arduino board are available which use other, more-powerful microcontrollers, have additional IO pins, and have different physical sizes. Devices operate at either a 5-V level and oscillator speed of 16 MHz or a 3.3-V level and 8 MHz. While many boards have an on-board USB connector to interface with a personal computer, the ATmega microcontroller communicates via a two-wire serial (transmit, Tx, and receive, Rx) connection. Boards with on-board USB connector also have a USB-serial converter chip and use a standard USB-USB cable, while other boards, to simplify design and lower cost, do not incorporate the USB-serial chip. A special cable, which contains the USB-serial chip and creates a virtual serial port, must be used.

The Arduino board is designed to allow expansion through the connection of auxiliary boards or shields. The shields connect via mating pins which are arranged in the same physical configuration as the Arduino board, and simply plug onto the headers on the top of the Arduino board. The shields are then controlled by the Arduino microcontroller and program, which access the shields' pins through the Arduino pins. Programming libraries allow users to quickly integrate new devices and sensors into projects without needing to write extensive new program routines.

Software

The software environment for programming and interacting with the Arduino board is available for download and installation for several computer operating systems (GNU/Linux, Mac OS X, and Windows). Using the IDE, the user writes programs in a language based on C++. The IDE then compiles and error-checks the program, and downloads the compiled routine to the microcontroller. A terminal window is available for outputting text and data from the Arduino board to the computer monitor and for interacting with the microcontroller.

As an open-source project, the Arduino benefits from the collective efforts and expertise of developers from around the world. Programming libraries, which contain routines to simplify programming and incorporate advanced features, sample code, and complete programs are available to download, use, and modify as needed. The IDE, libraries, and sample code can be accessed via the Arduino project website [6].

Communications

The Arduino development platform provides several methods of communicating with external components, sensors, and computers. In addition to built-in A/D converters and timers for measuring analog voltage signals, several standardized communications protocols are available for interfacing digital components and sensors.

The Inter-Integrated Circuit, also called I²C or I2C, protocol developed by Philips Semiconductor, is a twowire serial transfer protocol designed for communications between integrated-circuit chips and microcontrollers. Two IO pins on the Arduino's ATmega328 microcontroller are designated for I2C communication. Each I2C device has its own unique identification number and address, allowing multiple devices to be connected to the same I2C pins. The microcontroller initiates communication with a device by first sending the address of the device and then reading data from or writing data to the device. Identification numbers are unique to each type of component (memory chip, clock, temperature sensor, etc.) while addresses are either preset by the manufacturer or specified by the user through different hardware configurations.

The Dallas 1-Wire protocol, developed by Dallas Semiconductor, uses a single IO pin for communication and, optionally, to power the external 1-Wire device. Like I2C, multiple devices can be connected to a single 1-Wire pin, and are called by the microcontroller using the device's unique address.

The Serial Peripheral Interface, or SPI, is a four-wire system developed by Motorola and provides a serial data link that operates in full duplex mode. SPI devices communicate in master/slave mode using three IO pins, with the master device, the microcontroller, initiating communications with the slave, a sensor or other device. The microcontroller uses an additional IO pin for each device to select and communicate with a particular device.

RS-232 is the standard serial communication protocol that was widely used to communicate between personal computers and peripherals before the advent of the universal serial bus (USB). RS-232 uses two communication lines (Rx to receive, Tx to transmit), and is the protocol used by the Arduino' microcontroller to interface with a computer for programming. Since few modern computers contain an RS-232 port, a virtual serial port must be created. While some Arduino boards have a USB-to-serial converter chip on-board, many boards do not in order to reduce cost and power consumption. A special USBserial cable which contains the converter chip, such as the FTDI Cable (www.makerspace.com), interfaces to the computer's USB hub and creates a virtual serial port.

Sensors

A large number of sensors are available to monitor and measure many types of environmental parameters or physical processes. The rapid advances and usage of programmable microcontrollers have brought an increase in the availability and ease of use of sensing devices designed to interface with microcontrollers. The sensors operate at low voltages, and output signals compatible with microcontrollers, including analog voltages, varying frequencies, and a selection of digital communications protocols.

While the number of parameters sensed, and the number of sensors available, is vast, a few examples are presented and discussed in the following subsections.

Temperature

One of the most-common measurements made in a multitude of disciplines is temperature. A variety of temperature sensors is available using several different measurement technologies. While thermistors, which are sensors whose electrical resistance changes in response to temperature, are still in use, alternate electronic sensors are available which are designed to interface easily with microcontrollers and computers.

Analog temperature sensors, such as the LM35 (National Semiconductor, Santa Clara, CA USA) and TMP36 (Analog Devices, Inc, Norwood, MA USA), are designed to output a voltage signal proportional to temperature. The microcontroller supplies an excitation voltage to the sensor, and then measures the sensor's output voltage with an on-board A/D converter. The microcontroller program calculates temperature using a calibration developed by the sensor manufacturer. The LM35 sensor, for example, provides a linear response with a calibration of 10 mV/C: temperature (°C), is therefore calculated by dividing the output voltage, in mV, by 10. Analog sensors are usually very inexpensive and easy to work with, requiring only a simple voltage measurement and calibration equation to determine temperature. The microcontroller must have an A/D converter, and a stable reference voltage, which some may not have, requiring the addition of external components and circuitry.

Digital temperature sensors are designed to provide a calibrated and voltage-converted output which can be read directly as a temperature value. These sensors do not require a voltage measurement to be made, allowing the use of microcontrollers which do not have A/D converters. Digital sensors interface with the microcontroller through one of several communications protocols, such as I2C, 1-Wire, and SPI, with transfer of information accomplished via the microcontroller program. Digital sensors often have the feature of a unique identification number, allowing multiple sensors to be connected to the same IO pins on the microcontroller, thus not using additional pins. In contrast, since each analog sensor would require its own A/D input pin, multiple analog sensors could quickly fill available A/D converter pins.

For making non-contact temperature measurements, infrared thermometer (IRT) sensors are available which are inexpensive and easy to interface. The MLX90614 (Melexis SA, Ieper, Belgium) series of IRTs communicate with the microcontroller via the I2C protocol. Experience using these sensors to monitor crop canopy temperature [5] has shown them to work well in a harsh agricultural environment, operate for extended periods under battery power, and provide accurate temperature measurements.

Soil-Water Status

In many agricultural, natural-resource, and water-management disciplines, water availability and moisture status are of great importance. The amount of water available in the soil profile for extraction by growing plants can be measured with a water-content sensor. A water potential sensor provides a measure of how tightly the water is held to the soil particles and how much energy must be expended to extract the water by the plant roots. This can be related to the availability of water to the plant.

Many of the currently available water-content sensors rely on a measure of the capacitance of the soil-water environment. Dielectric properties of the soil-water system vary weakly with soil properties, such as mineral composition, bulk density, and organic-matter content, but are strongly influenced by water content [13]. Watercontent sensors, such as the EC-5 and EC-20 (Decagon Devices, Pullman, WA USA), and VG400 (Vegetronix, Bluffdale, UT USA), consist of a capacitive-sensing element and on-board electronic circuitry. When powered by the microcontroller, the sensors return a voltage signal proportional to the water content in the soil. Measuring the voltage with the microcontroller's A/D converter and applying a calibration equation in the microcontroller program results in a water-content value, expressed in units of volume of water/volume of soil. Sensor manufacturers may provide calibration equations for limited soil types and other porous media, such as potting soil or greenhouse media, but the user often must develop a calibration, or at least verify the manufacturer's, under his specific soil conditions to obtain accurate water-content measurements.

Water-potential sensors are usually designed to act as variable resistors, in which the electrical resistance of the sensor varies in response to its water content. The sensor is composed of a porous matrix, and water can move into and out of the matrix in response to the matric potential of the soil. As the water content in the porous matrix changes with matric potential, the electrical resistance also changes. A calibration equation then converts resistance to matric or water potential, expressed in units of kiloPascals (kPa).

The Watermark 200SS (Irrometer Company, Riverside, CA USA) water-potential sensor is popular in irrigation-scheduling applications due to its ease of installation and low cost. It requires an alternating-current excitation rather than direct current, however, which can involve additional care when interfacing with a microcontroller (see [3] and the discussion in Section 3.1 below for alternative implementations). To allow direct connection and use with any microcontroller circuit, the MPS-2 (Decagon Devices, Pullman, WA USA) is designed to operate from a direct-current supply and output a simple voltage signal in response to soil-water potential. The voltage signal is measured with the microcontroller's A/D converter and then converted to water potential with a calibration equation.

Distance/Height

Distance measurements are common in robotic and industrial/manufacturing environments to determine distance from a moving vehicle for obstacle avoidance, detect presence or absence of material, and ensure proper placement of a component. In research applications, distance measurements can be used to determine properties such as plant height and canopy width, depth of water in canals, and fluid levels in tanks.

Distance measurements are commonly made using two sensing technologies, ultrasonic and infrared. Ultrasonic sensors often consist of two transducers, one which emits a pulse of high-frequency sound waves, and a second one to detect the sound after reflecting off a nearby surface. Distance is determined by measuring the length of time between sending the pulse and receiving the reflection, or echo, and converting this to a distance based on the speed of sound. Ultrasonic sensors, such as the SRF series (Devantech Ltd., Norfolk, UK) and the PING (Parallax Inc., Rocklin, CA USA) interface with a microcontroller via one or two digital IO pins. The microcontroller is programmed to initiate a pulse, then starts an internal timer and counts the number of microseconds until an echo signal is detected, and calculates the distance based on this time interval. Sensors are available with varying fields of view to enable sensing over wider or narrower regions.

Infrared sensors operate by emitting a beam of light and detecting the reflected beam, after hitting an obstacle, with a light sensor. The reflected beam returns at a slight angle from the emitted beam, and the angle of the two beams is dependent on the distance of the obstacle from the sensor. The reflected beam strikes the light sensor at some point, and is read by an on-board microcontroller which is programmed to output an analog voltage in proportion to distance. The analog voltage is input to the Arduino microcontroller's A/D converter and converted to distance with a calibration equation supplied by the manufacturer. Infrared sensors such as the GP2 series (Sharp Electronics Corporation, Mahwah, NJ USA) offer a variety of operating ranges.

Pressure

Maintaining proper pressure and measuring the existing pressure are important in many processes and environments. Atmospheric air pressure is an important meteorological parameter, for example, and liquid pressure can be used to determine fluid depth based on hydrostatic pressure relationships.

Many pressure sensing devices are available and range from simple sensing elements to amplified, calibrated, and temperature-compensated sensors. Sensing configurations typically consist of piezoresistive elements and a silicon diaphragm arranged in a Wheatstone-bridge circuit. A change in pressure causes the diaphragm to flex and changes the resistance values of the piezoresistive elements. Since changes are very small, the change in electrical output of the Wheatstone bridge is also small, requiring accurate voltage-measuring circuitry. Amplifying the output signal allows the signal to be measured with an A/D converter on the Arduino. Temperature changes can also affect the piezoresistive elements, resulting in the need for temperature compensation under conditions of large temperature swings. A range of pressure sensors, including the non-temperature-compensated 24PC, temperature-compensated 26PC, and fully compensated and amplified 40PC series (Honeywell Sensing and Control, Golden Valley, MN USA) can be interfaced and read with the Arduino's microcontroller.

Resolution of Analog Sensor Measurements

Analog sensors output a voltage signal which is converted into a numerical value by an A/D converter. The A/D converter is characterized by a known, reference voltage, which determines the range of acceptable voltage signals, and the number of digital values, or bits, into which the voltage range is divided. The Arduino's microcontroller contains a 10-bit A/D converter, meaning that the voltage range is divided into 2^{10}, or 1024, divisions. To measure a sensor's voltage signal, the A/D converter compares the voltage level to the reference voltage, and returns a proportional digital value in the range of 0 to 1023.

The A/D converter characteristics determine the resolution and accuracy of voltage measurements. The resolution, or smallest change in voltage that the A/D converter can detect, is dependent on the A/D converter's number of bits and the reference voltage. The Arduino's microcontroller has a built-in 1.1 V reference, which provides the A/D converter with a resolution of 1.1 V/1024 bits, or 0.00107 V/bit. The microcontroller's 5-V power supply voltage can also be used as a reference, resulting in an A/D conversion resolution of 0.00488 V/bit.

Resolution can be increased or decreased by changing the number of A/D conversion bits. External A/D converter chips are available which have higher-bit resolutions and can be easily interfaced with the Arduino. The MCP3424 (Microchip Technology Inc., Chandler, AZ USA) is an A/D converter chip which can read four input voltage signals with 18-bit (262,144 divisions) resolution. With a 5-V reference voltage, this would provide an A/D resolution of 0.0000191 V/bit. The MCP3424 communicates with the microcontroller using I2C.

To illustrate the effect of A/D converter resolution on sensor measurements, consider an analog temperature sensor that outputs a voltage signal between 0 and 5 V over a temperature range of 0 to 65 C. The signal, therefore, changes by 65 C/5 V, or 13 C/V. Using the microcontroller's built-in 10-bit A/D converter and a 5-V reference, with a resolution of 0.00488 V/bit, the resolution of temperature measurements would be 13 C/V*0.00488 V/bit, or 0.06 C/bit, which would be acceptable for most applications.

The resolution of a signal from a non-amplified pressure sensor, with an output of 0 to 10 mV over a range of 0 to 100 kPa, would have a measurement resolution of 100 kPa/0.01 V*0.00488 V/bit, or 48.8 kPa/bit. This would be unacceptable, providing only three measurements (0, 48.8, and 97.6 kPa) over the entire measurement range. Using the MCP3424 external A/D converter, with 18-bit resolution, would greatly improve voltagemeasurement capability and provide a pressure-measurement resolution of 0.038 kPa/bit.

Time-Keeping

In many data-collection efforts, proper timing of measurements and dateand time-stamping of sensor data are required. The microcontroller on the Arduino board has a very accurate 16 MHz oscillator and the ability to measure time increments with microsecond accuracy, but is not designed to provide real time (hours, minutes) and date information. If electrical power to the microcontroller is lost, the oscillator and microcontroller program cease to function, and any timing information is also lost.

External real-time clock (RTC) chips are used to provide time-keeping functions, with dedicated built-in or added backup batteries to retain accurate time information. RTCs such as the DS1307 and DS1337 (Maxim Integrated Products, Inc., Sunnyvale, CA USA) interface with the microcontroller using the I2C protocol, while others, such as the MCP795 (Microchip Technology Inc., Chandler, AZ USA) communicate via SPI. Simple routines in the microcontroller program access the RTCs to set or read time and date information, which can then be used to trigger sensor measurements at regular time intervals or record timing information of events.

Data Storage

Data collection often involves long-term, automated storage of sensor measurements. While the Arduino's microcontroller has extensive memory available for program storage, non-volatile data-storage capability is limited. On-board memory consists of 1 kb (1000 bytes), so a maximum of 1000 data values could be stored and retained if battery power were interrupted. To expand the storage capacity, external storage must be added.

External memory chips are available with varying amounts of non-volatile memory. The 24LC family of memory chips (Microchip Technology Inc., Chandler, AZ USA), for example, are available in capacities from 16 bytes to 65,356 KB. These chips communicate via the I2C protocol and have individual identification numbers so that multiple chips could be connected to increase storage amounts considerably.

For permanent or large-capacity storage, add-on boards are available which provide data storage to standard SD memory cards (Adafruit Industries, New York, NY USA) or microSD memory cards (Sparkfun Electronics, Boulder, CO USA). Memory cards are commonly available with storage capacities from 1 gigabyte (GB) to several GB, are inexpensive, and can be easily interfaced with the Arduino hardware. Since the memory cards can be read with a computer, data can be transferred quickly and easily between datalogger and computer. Software libraries have been written to provide all memory card reading, writing, and data-access functions, enabling rapid incorporation of memory-card storage into a datalogging project.

SENSING APPLICATIONS

To illustrate how the Arduino platform can be used to develop and implement an inexpensive, automated data collection and monitoring program, several examples are presented. These examples include a brief description of the circuitry and details of the project implementation. Microcontroller programs are not included but are freely available by contacting the authors.

Soil-Moisture Monitoring Datalogger

Monitoring moisture status of the soil profile is useful in scheduling irrigations and monitoring the movement or availability of water in the soil profile. Sensors are installed in the soil profile at various depths within a crop's root zone and are monitored periodically. A datalogger was designed to record measurements from three soilmoisture sensors at one-hour intervals, and store the measurements, along with the date and time, to a microSD memory card.

Hardware

The main components of the datalogger include an Arduino-compatible microcontroller board, voltage regulator, microSD/prototyping shield, and real-time clock/ calendar. The Diavolino microcontroller board (Evil Mad Science LLC, Sunnyvale, CA USA) was chosen for its low cost, simple and low-component design, and ease of modification for battery-powered operation. The board, designed to operate from the 5 V power supplied via the USB computer connection, was modified by adding a two-pin header to connect an external AA battery pack. An LP2950 voltage regulator (National Semiconductor Corp., Santa Clara, CA USA) and capacitors were added to convert the unregulated battery voltage to a stable 5-V source to power the microcontroller. A trace on the printed circuit board, which powered the board from the USB connection, was then cut so that the only power source was the AA battery pack. The modified Diavolino microcontroller board is shown in **Figure 1**.

The microSD shield (Sparkfun Electronics, Boulder, CO USA) consists of a microSD-card holder, with on-board voltage-level shifter to supply the proper voltage levels for reading from and writing to a microSD card, and a prototyping area to incorporate additional circuitry into the shield. The microSD shield was designed to be powered from the microcontroller board's power supply, thus the microSD card and voltage-level shifter would always be powered and continuously drawing current. The shield was modified for battery-powered operation by rerouting the power supply for the microSD card and voltage-level shifter to one of the microcontroller's digital pins so that the components could be turned on and off as needed. A microSD card (Samsung) with a 2 gigabyte storage capacity was then inserted into the microSD card holder.

A circuit was designed and added to the microSD shield's prototyping area to measure the output from three soil-moisture sensors. A DS1337 real-time clock/ calendar chip provides date and time information for the microcontroller to make sensor readings at regular time intervals and to dateand time-stamp sensor data stored to the microSD card. A 32.768 kHz crystal oscillator provides an accurate timing signal for the DS1337, and a 3.3-V lithium coin cell battery powers the clock chip. The DS1337 interfaces with the microcontroller via the I2C protocol.

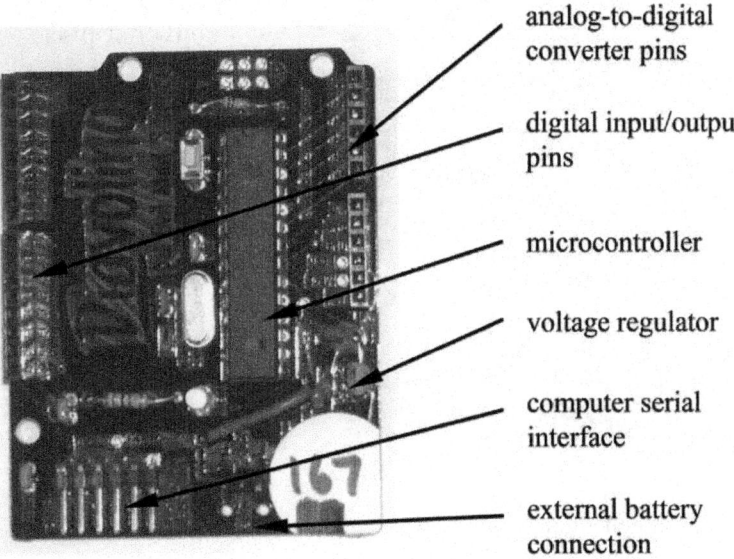

analog-to-digital converter pins

digital input/output pins

microcontroller

voltage regulator

computer serial interface

external battery connection

Figure 1: Modified Diavolino Arduino-compatible microcontroller board.

The soil-moisture sensors consist of three Watermark 200SS matric-potential sensors whose electrical resistance varies with moisture content. A circuit was designed in which each sensor, which acts as a variable

resistor, forms one leg of a half bridge, or voltage divider. The half bridge is connected to two digital pins on the microcontroller, and each voltage-divider output is connected to an A/D pin. A photograph of the completed circuit, mounted on the microSD shield, is shown in **Figure 2**, and a schematic of the circuit is shown in Figure 3. A list of materials, with sources and approximate cost (small-quantity retail price, in the United States, US dollars, 2011), is provided in **Table 1**.

Software

Using the Arduino IDE installed on a personal computer, a microcontroller program, called a sketch on the Arduino platform, was written to read the real-time clock, make soil-moisture sensor measurements, and store the time and sensor data to a microSD card. Communication between the computer and microcontroller board requires an RS-232 serial connection, which was accomplished via an FTDI USB-serial cable, which interfaces to the computer's USB hub and creates a virtual serial port.

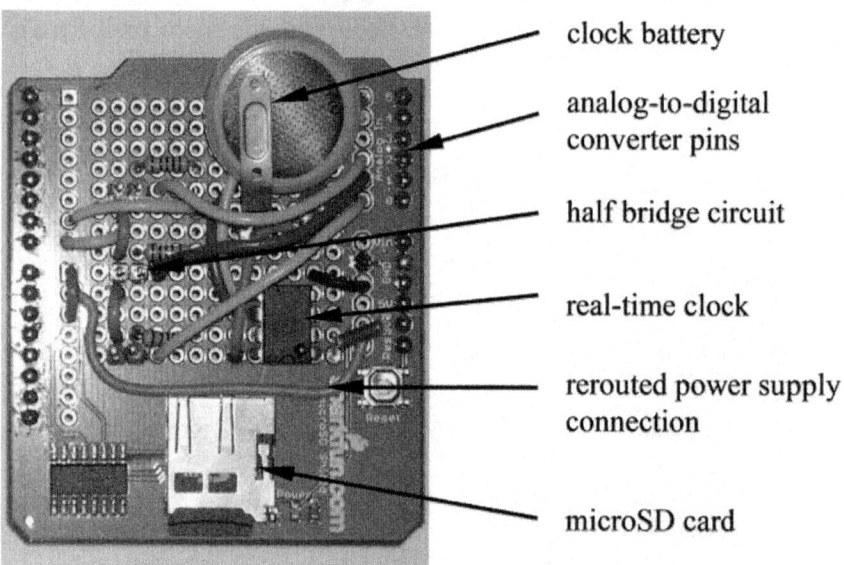

clock battery

analog-to-digital converter pins

half bridge circuit

real-time clock

rerouted power supply connection

microSD card

Figure 2: Modified microSD shield with circuit components installed.

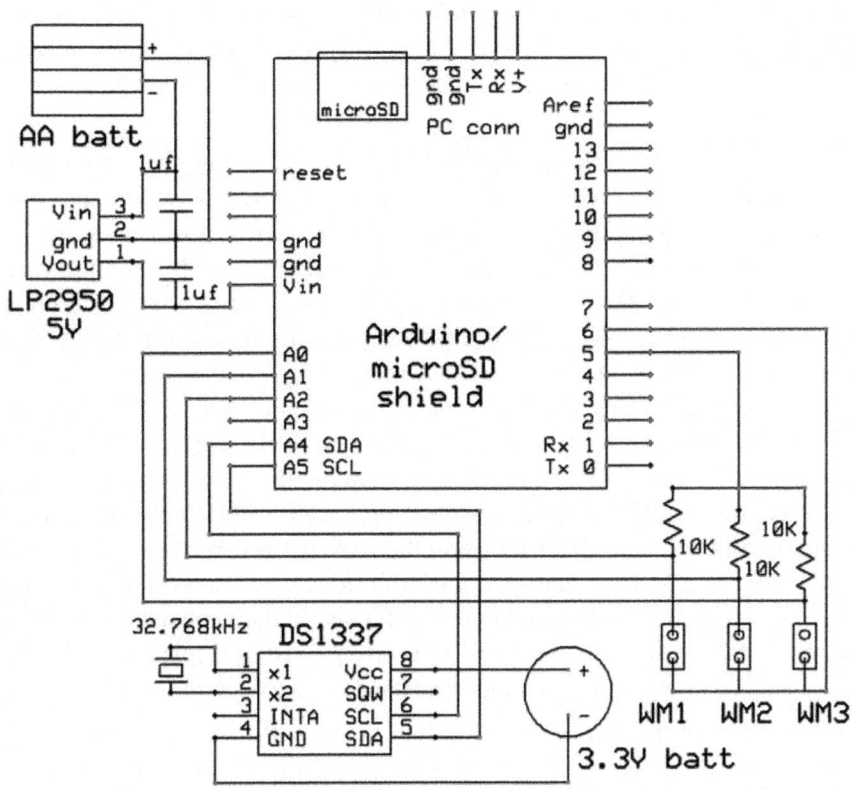

Figure 3: Schematic for soil-moisture sensor datalogger.

Table 1: List of materials for soil-moisture sensor datalogger

Main components	Part number	Supplier	Cost $
Microcontroller board	Diavolino	Evil Mad Science	13
microSD/prototyping shield	microSD shield	Sparkfun Electronics	14
microSD card	2 gb	Samsung	4
Real-time clock/calendar	DS1337	Maxim Integrated Products	3
Oscillator	32.768 kHz	Citizen America	1
Regulator	LP2950 5V	National Semiconductor	2
	Miscellaneous (capacitors, resistors, headers, batteries)		7
	Datalogger Total		44
Soil-moisture sensor	200SS	Irrometer	30

To enable long-term, battery-powered operation of the datalogger, the microcontroller was programmed to spend most of its time in a low-power, sleep mode. Periodically, the microcontroller would wake up and read the current time from the real-time clock. If it was time to take a measurement, the microcontroller would power the measurement circuit, otherwise it would go back to sleep. At one-hour intervals, the measurement circuit on the microSD shield was enabled, and the soil-moisture sensors were read and data stored to the microSD card.

To properly read a Watermark 200SS sensor, an alternating current source is recommended in order to avoid polarizing the sensor with a prolonged direct-current excitation, which can influence sensor measurements and degrade the sensor over time. The microcontroller can only supply a direct-current excitation, however, so a pseudo-alternating current source was created by rapidly switching the polarity of the direct-current voltage sent to power the sensor, and the sensors were then read under each polarity. Digital pin 6 was first set high (a voltage level of 5 V) and pin 5 was set low (a voltage level of 0 V) so that current flowed through the half bridge in one direction (see**Figure 3**). The output voltage, Vout, between the 10 kohm resister, R, and the Watermark sensor, Rwm, was measured with an A/D converter, and the sensor resistance was calculated using the voltage-divider relationship, Vout = R/(Rwm + R)*5 V. The polarity of the half bridge was then switched by setting pin 5 high and pin 6 low, so that current flowed in the opposite direction, and output voltage was again measured and sensor resistance calculated. This was repeated five times, and an average resistance was calculated.

To arrive at the sensor's final output, namely the matric-potential of the soil, in kPa, a calibration equation is required to convert sensor resistance to matric potential. Much work has been done calibrating and verifying the Watermark 200SS sensor [14-16], and several calibration equations have been proposed. The equation of Shock et al. [16] was chosen, written as SWP = (4.093 + 3.213 Rwm)/(1 − 0.009733*Rwm − 0.01205*Tsoil), where SWP is the soil-water potential (kPa), Rwm is the sensor resistance (ohms), and Tsoil is the soil temperature (°C). While sensor performance has been shown to vary slightly with temperature, and a temperature-correction factor is included in the calibration equation, soil temperature was not measured and, instead, a constant temperature of 25 °C was used. To improve accuracy of sensor readings, a soil-temperature sensor could be added to the datalogger circuit and actual temperature measurements input to the calibration equation.

Following sensor measurements, power was sent to the microSD card circuit, and the data were stored to the microSD card. Data were stored as ASCII text, separated by spaces, in a plain text file, and consisted of six values; a datalogger board identification number, date (month/ day/year), time of day (hour), sensor #1 reading (kPa), sensor #2 reading (kPa), and sensor #3 reading (kPa). The microcontroller then turned all power off to the microSD shield and returned to low-power, sleep mode.

Data

Thirty soil-moisture sensor dataloggers were constructed and deployed to monitor soil-moisture status in experimental research plots at the USDA Agricultural Research Service's Jamie Whitten Delta States Research Center at Stoneville, Mississippi USA. Research plots planted to soybean and cotton were instrumented with soil-moisture sensors and Arduino-based dataloggers. At each instrumented site, Watermark sensors were installed at three depths; 15-, 30-, and 60-cm, below the soil surface. The sensors were connected to a datalogger, and the datalogger was placed inside a weatherproof plastic enclosure attached to a wooden stake driven into the ground. The datalogger was turned on, and collected sensor data at one-hour intervals throughout the entire growing season. Periodically, each site was visited to download data from the microSD card to a portable tablet computer. The text data files were then returned to the office, uploaded to a desktop computer, and input to a spreadsheet for analysis and viewing. Typical data, from one site over a sevenweek period following planting in 2011, are shown in **Figure 4**.

Soil-water potential values near 0 indicate very moist soil conditions, with soil-water levels decreasing as the water-potential values become more negative. Hourly data from the three sensors were input to a spreadsheet, and the average of the three sensor readings was calculated. The average values were used to determine when an irrigation was needed. When the average values reached a threshold value of −50 kPa, an irrigation was scheduled. In **Figure 4**, soil-water levels decreased early in the season as the growing crop extracted water until rainfall occurred on 6/21, rewetting the soil. As soil water was used by the crop, the levels dropped until reaching −50 kPa, and two irrigations were required. Evident in the data are differences in water use with depth in the soil profile. Early in the season, changes in water potential were slower at 30 cm than at 15 cm, and much slower at 60 cm, suggesting more active roots in the shallower depths. As the season progressed, water-use rates increased at the 30-cm depth, and later at the 60-cm depth, suggesting increases in root activity and water extraction.

Ultrasonic Water-Level Datalogger

Fluid levels are measured in a variety of applications; fuel tanks, water reservoirs, and irrigation canals, for example. Evaporation pans are used to estimate the evaporative demands of the atmosphere in order to determine crop water use and soil evaporation rates for input in water-balance and evapotranspiration studies, and to assist in irrigation scheduling. A datalogger was de-

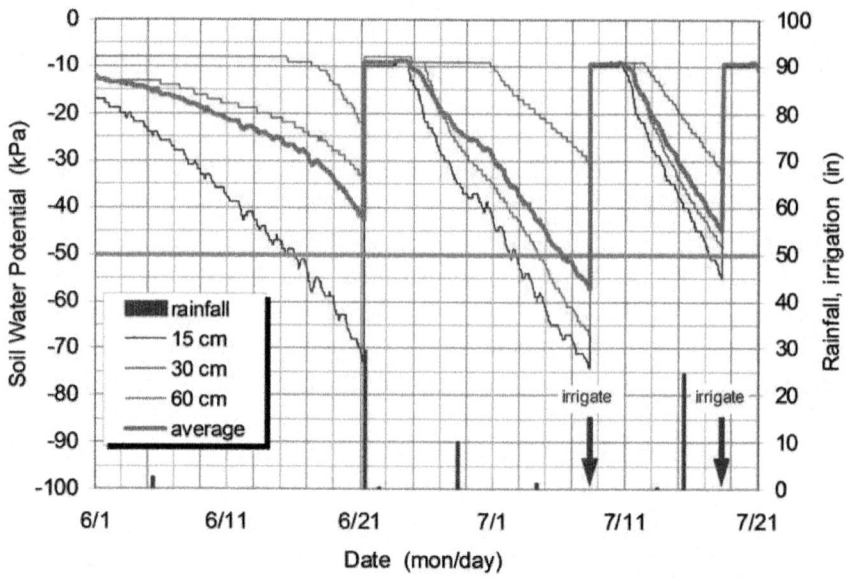

Figure 4: Data collected with the soil-moisture datalogger for a seven-week period in 2011.

veloped to automate the measurement of the depth of water in an evaporation pan using an ultrasonic distance sensor.

Hardware

The ultrasonic water-level datalogger circuit is based on that of the soil-moisture datalogger, and incorporates many of the same circuit components. The same Ardinocompatible microcontroller board was used, and was modified in the same manner to supply a stable power source and enable battery-powered operation. The same microSD/prototyping shield and real-time clock components were also used. A schematic of the ultrasonic water level datalogger is shown in **Figure 5**.

An ultrasonic distance sensor, model SRF-04 (Devantech Ltd., Norfolk, UK), interfaces with the microcontroller via three digital pins; power, trigger, and echo pulse. The sensor consists of two ultrasonic transducers, one to send an ultrasonic pulse and one to receive the pulse's echo. To make a measurement, power is supplied to the sensor, and a measurement is initiated by sending a brief signal to the trigger pin, which causes an ultrasonic pulse to be sent. The microcontroller then begins monitoring the echo pulse pin, and measures the length of time it takes to receive an echo signal.

A temperature sensor was added to measure the air temperature of the environment. The LM35 analog temperature sensor outputs an analog-voltage signal in proportion to its temperature. The signal is input to one of the microcontroller's A/D converters, and a calibration equation supplied by the manufacturer is used to convert the voltage signal to temperature.

Software

The microcontroller program for the ultrasonic waterlevel datalogger used many of the same routines written for the soil-moisture datalogger. The microcontroller

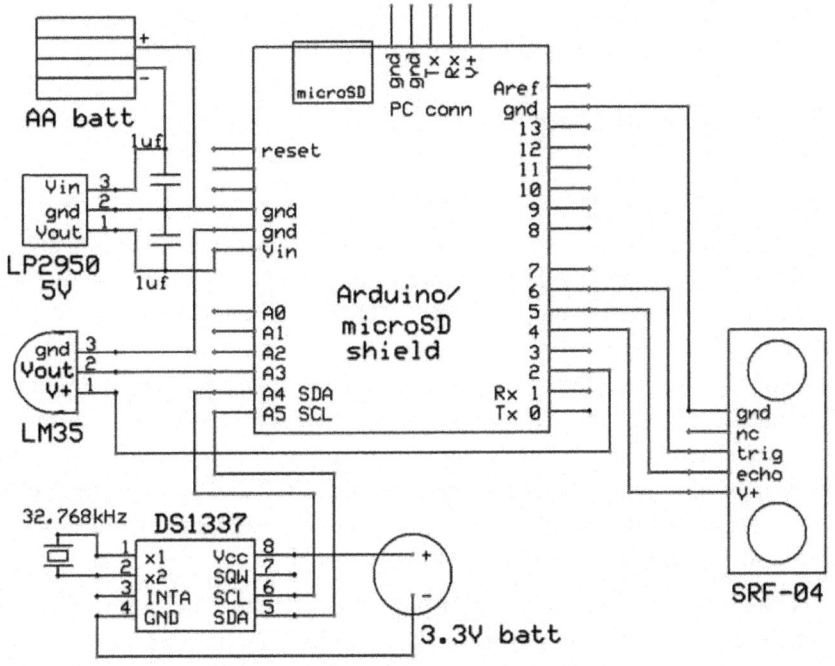

Figure 5: Schematic for ultrasonic water-level datalogger.

wakes periodically from a low-power sleep mode to read the real-time clock and determine if it is time to take measurements. If so, measurements are taken and recorded, otherwise, the microcontroller goes back to sleep.

At each measurement interval, the microcontroller turns on the temperature sensor and makes an air temperature measurement. The ultrasonic sensor is then powered, a trigger signal is sent, and the time for an echo signal to return is measured. The time for the echo to return is then used to calculate the distance, based on the speed of sound, between the sensor and the surface upon which the ultrasonic pulse impacted. The speed of sound, however, is strongly dependent on the air temperature, and slightly affected by humidity [17], and can be corrected to improve the accuracy of distance measurements using the relationship $v = 331$ m/s $+ 0.6$ m/s/C*T, where v is the speed of sound (m/s) and T is the air temperature (°C). To make a distance measurement, the air temperature measurement is first used to correct the speed of sound value. The speed of sound and the time taken to return the pulse echo are then used to calculate the distance from the sensor to the reflecting surface. This distance is then subtracted from the distance of the sensor to the bottom of the evaporation pan, measured previously when installing the ultrasonic sensor, to determine the depth of water in the pan.

Following air temperature and water level measurements, the data, microcontroller board identification number, and date and time are written to the microSD card. The microcontroller then turns off power to the circuit and returns to low-power, sleep mode.

Data

Two ultrasonic water-level dataloggers were constructed and installed in summer 2011 and operated for a threemonth period. The sensors were installed on an evaporation pan approximately 300 mm above the bottom of the pan. Sensor measurements were recorded at one-hour intervals, and the data were periodically downloaded from the microSD card during periodic site visits. During site visits, manual measurements of the water level were made by inserting a steel ruler into the evaporation pan and reading the depth of water. The depth of water in the pan varied between 70 and 195 mm, decreasing as water evaporated in response to the environmental demand and increasing due to rainfall and periodic manual refilling.

Data collected during a four-day period with one ultrasonic sensor datalogger are shown in **Figure 6**. Data include air temperature, raw depth (before correcting the speed of sound for temperature) and temperature-corrected depth, and manual measurements of the water levels. Large increases in apparent depth of water can be seen in the raw sensor readings each morning beginning around 6:00, as the sun rose and air temperature increased rapidly. The raw depths also continued to appear to decrease after sunset, when evaporation would be expected to cease. Correcting the speed of sound for air temperature mostly eliminates these errors, resulting in expected changes in water level, decreasing during daylight hours and minimal changes during nighttime. An increase in depth can be seen in response to a manual addition of water to the pan.

Accuracy of ultrasonic measurements was determined by comparing water levels measured with the ultrasonic sensors to those measured manually. Manual depth measurements were made 18 times, at varying times throughout the three-month period and at varying times of day. Manually measured water levels ranged from 75 to 158 mm. Comparison of measurements from the two ultrasonic sensors is shown in **Figure 7**, and indicates a very good agreement with the manual measurements, with a standard error of measurements of approximately 2 mm.

Environmental Datalogger

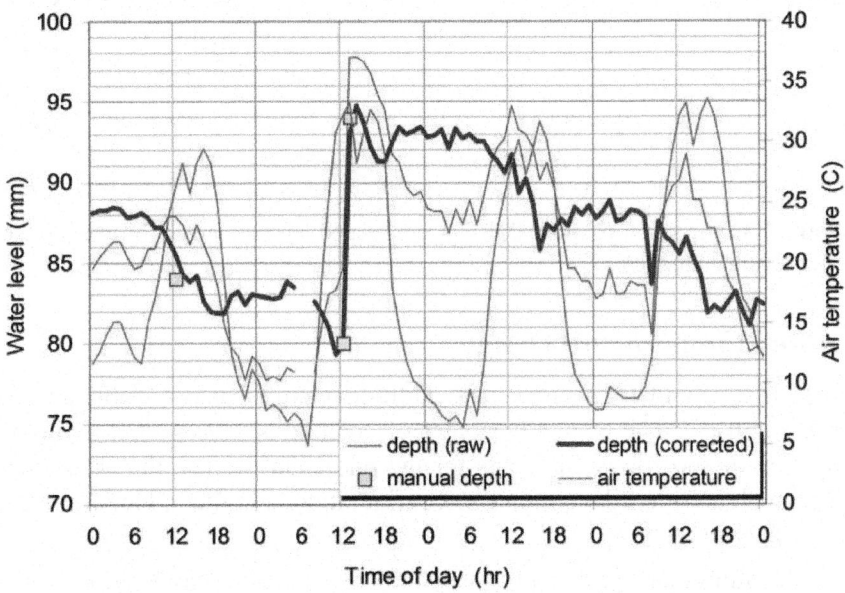

Figure 6. Hourly data collected with one ultrasonic water-level datalogger during a four-day period.

An Arduino-based datalogger can also be built using a custom printed circuit board (PCB) rather than starting with a commercially available board. A datalogger was designed and fabricated to collect environmental data in a forested setting. The datalogger was de signed to accommodate a variety of sensor types, but was primarily intended to measure soil moisture and air temperature. Rather than developing a system around a commercially available Arduino board, a custom PCB was created which contained the Arduino microcontroller and other components.

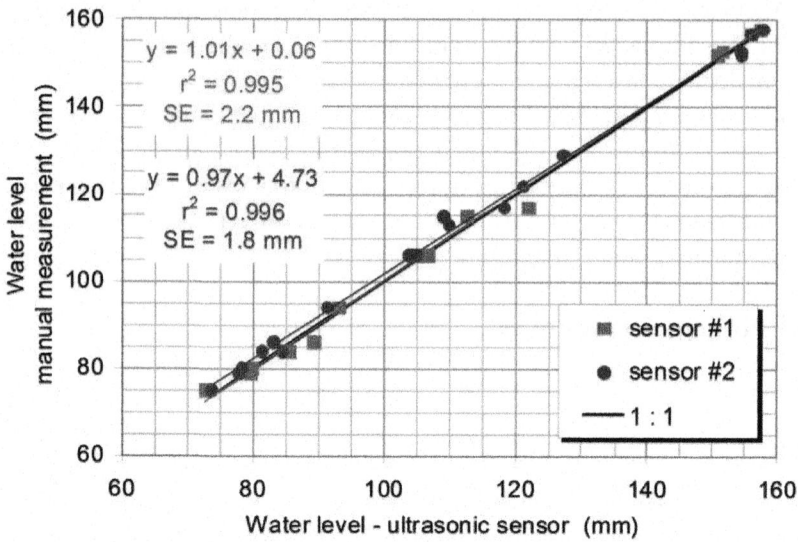

Figure 7: Comparison of manual versus automated measurements for two ultrasonic water-level dataloggers.

Custom Circuit-Board Design

Creating a new PCB requires some additional skills but is a viable approach for many users. The main advantages of creating a custom PCB include the selection of specific components for the particular project, the creation of a board with a custom layout or size/configuration, and the reduction, in some cases, in total cost.

The minimum components needed to create an Arduino board include the ATmega328 microcontroller, a resonator (self-contained oscillator circuit), a reset button, a voltage regulator, connectors for a battery pack, computer interface, and a few resistors and capacitors. ATmega328 microcontroller chips are available preprogrammed with the Arduino-system's bootloader, enabling the use of the Arduino IDE to create and upload programs to the microcontroller. Additional components for most datalogger projects include a real-time clock/ calendar, a memory device, and one or more light-emitting diodes (LEDs) to indicate the operational status of the datalogger. All of these components are readily available as through-hole components which can be soldered to the PCB with a soldering iron.

The process of creating a custom circuit board begins with circuit and PCB design. Several software packages are available, some in freely available, open-source versions, to design the electrical schematic and then lay out the circuit on a PCB. A graphical user interface simplifies design, and the software creates a set of files in formats standardized for PCB manufacturing, which can then be transmitted to a PCB manufacturer. The manufacturer produces the bare PCB, and the final board is constructed by soldering the components to the PCB by hand.

Hardware

A board was designed using the freely available Design-Spark PCB software (www.designsp ark.com/pcb). The circuit was designed using a graphical schematic view, in which connections between circuit components are created but the actual size, shape, and layout of the components are unimportant. This schematic is then transferred to a printed circuit board layout, where the software suggests the physical layout and connecting traces of the components. The user is able to modify the layout as desired, to create a PCB that is easy to assemble, or which fits certain dimensional or other constraints.

The resulting board design, with dimensions of approximately 60 × 90 mm, was then electronically transmitted for fabrication using SeeedStudio's Fusion PCB service (www.seeedstudio.com/propa gate). Dataloggers were assembled by soldering circuit components to the custom PCB, with each datalogger requiring approximately 20 minutes to complete. A list, with approximate cost of components, excluding sensors, is provided in **Table 2**. The original design layout is shown in**Figure 8**(a), with resulting bare printed circuit board and finished datalogger board shown in Figures 8(b) and 8(c), respectively.

Dataloggers were deployed in the field along with air-temperature and soil-moisture sensors. Air temperature was measured using a DS18B20 12-bit digital temperature sensor (Maxim Integrated Products, Inc., Sunnyvale, CA USA). The sensor uses the 1-Wire communication protocol to transfer measurements to the microcontroller, and contains an internal 18-bit A/D converter which provides temperature measurements with a resolution of 0.06 C. Soil-moisture measurements were made using an EC-20 capacitive sensor. The microcontroller provided an excitation voltage to power the sensor via a digital IO pin, and measured the analog output voltage with a built-in A/D converter.

A battery pack consisting of 5 AA alkaline batteries enabled long-term remote operation by ensuring adequate voltage as the batteries discharged.

SUMMARY

Advances in electronic technologies, microcontrollers, and sensors offer researchers a variety of new and inex

Table 2: List of materials for environmental datalogger

Main components	Part number	Supplier	Cost $
Microcontroller with bootloader	ATmega328	Sparkfun Electronics	6
Printed circuit board		SeeedStudios	3
Memory chip	24LC512	Microchip	4
Real-time clock/calendar	DS1307	Maxim Integrated Products	1
Oscillator	32.768 kHz	Citizen America	1
Regulator	LP2950	National Semiconductor	1
Screw terminals			4
Miscellaneous (capacitors, resistors, connectors, LEDs)			6
Datalogger Total			26

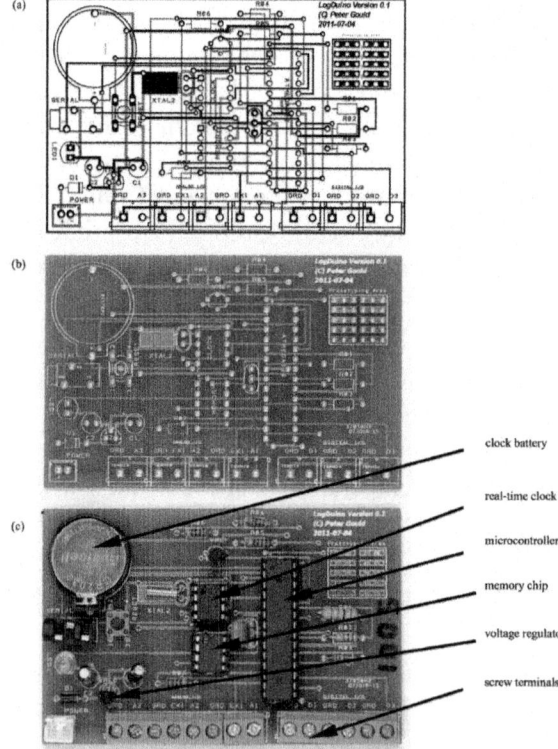

Figure 8: Arduino-based datalogger fabricated on a custom printed circuit board: top, (a) circuit-board layout; middle; (b) bare circuit board; bottom; (c) completed board with components installed.

pensive sensing, monitoring, and control capabilities The concept of open-source hardware, in which hardware designs, software programs, and development efforts are made freely available to all, help facilitate and expand the adoption of these capabilities. The open-source hardware Arduino development platform has great potential for implementation in scientific research applications, and can empower researchers with flexible, inexpensive tools for expanding their data-collection, automation, and control capabilities.

DISCLAIMER

Mention of a trade name, proprietary product, or specific equipment does not constitute a guarantee or warranty by the United States Department of Agriculture, and does not imply approval of the product to the exclusion of others that may be available.

REFERENCES

1. F. H. Moody, J. B. Wilkerson, W. E. Hart and N. D. Sewell, "A Digital Event Recorder for Mapping Field Operations," Applied Engineering in Agriculture, Vol. 20, No. 1, 2004, pp. 119-128.

2. K. A. Noordin, C. C. Onn and M. F. Ismail, "A Low-Cost Microcontroller-Based Weather Monitoring System," CMU Journal, Vol. 5, No. 1, 2006, pp. 33-39.

3. D. K. Fisher, "Automated Collection of Soil-Moisture Data with a Low-Cost Microcontroller Circuit," Applied Engineering in Agriculture, Vol. 23, No. 4, 2007, pp. 493-500.

4. G. Vellidis, M. Tucker, C. Perry, C. Kvien and C. Bednarz, "A Real-Time Wireless Smart Sensor Array for Scheduling Irrigation," Computers and Electronics in Agriculture, Vol. 61, No. 1, 2008, pp. 44-50. doi:10.1016/j.compag.2007.05.009

5. D. K. Fisher and H. Kebede, "A Low-Cost Microcontroller-Based System to Monitor Crop Temperature and Water Status," Computers and Electronics in Agriculture, Vol. 74, No. 1, 2010, pp. 168-173. doi:10.1016/j.compag.2010.07.006

6. Arduino, "An Open-Source Electronics Prototyping Platform," 2012. http://www.arduino.cc

7. D. Bri, H. Coll, M. Garcia and J. Lloret, "A Multisensor Proposal for Wireless Sensor Networks," 2nd International Conference on Sensor Technologies and Applications, Cap Esterel, 25-31 August 2008, pp. 270-275.

8. L. Buechley and M. Eisenberg, "The LilyPad Arduino: Toward Wearable Engineering for Everyone," Pervasive Computing, Vol. 7, No. 2, 2008, pp. 12-15.doi:10.1109/MPRV.2008.38

9. J. Zhang, S. K. Ong and A. Y. C. Nee, "Design and Development of a Navigation Assistance System for Visually Impaired Individuals," Proceedings of the 3rd International Convention on Rehabilitation Engineering & Assistive Technology, Singapore, 22-26 April 2009.

10. N. W. Bergmann, M. Wallace and E. Calia, "Low Cost Prototyping System for Sensor Networks," 6th International Conference on Intelligent Sensors, Sensor Networks and Information Processing, Brisbane, 7-10 December 2010, pp. 19-24.doi:10.1109/ISSNIP.2010.5706802

11. D. Gordon, M. Beigl and M. A. Neumann, "Dinam: A Wireless Sensor Network Concept and Platform for Rapid Development," 7th International Conference on Networked Sensing Systems (INSS), Kassel, 15-18 June 2010, pp. 57-60. doi:10.1109/INSS.2010.5573290

12. J. Sarik and I. Kymissis, "Lab Kits Using the Arduino Prototyping Platform," Frontiers in Education Conference, Washington DC, 27-30 October 2010, pp. 1-5.

13. A. M. Thomas, "In situ Measurement of Moisture in Soil and Similar Substances by 'Fringe' Capacitance," Journal of Scientific Instrumentation, Vol. 43, No. 1, 1966, pp. 21-27.doi:10.1088/0950-7671/43/1/306

14. S. J. Thomson and C. F. Armstrong, "Calibration of the Watermark Model 200 Soil Moisture Sensor," Applied Engineering in Agriculture, Vol. 3, No. 2, 1987, pp. 186- 189.

15. E. P. Eldredge, C. C. Shock and T. D. Stieber, "Calibration of Granular Matrix Sensors for Irrigation Management," Agronomy Journal, Vol. 85, No. 6, 1993, pp. 1228-1232.doi:10.2134/agronj1993.000219620085000 60025x

16. C. C. Shock, J. M. Barnum and M. Seddigh, "Calibration of Watermark Soil Moisture Sensors for Irrigation Management," Proceedings of the International Irrigation Show, San Diego, 1-3 November 1998, pp. 139-146.

17. D. A. Bohn, "Environmental Effects on the Speed of Sound," Journal of the Audio Engineering Society, Vol. 36, No. 4, 1988, pp. 223-231.

Chapter 6

APPLYING SOFTWARE ENGINEERING METHODOLOGY FOR DESIGNING BIOMEDICAL SOFTWARE DEVOTED TO ELECTRONIC INSTRUMENTATION

[1]Gilsa Aparecida de Lima Machado, [1] Patricia Mara Danella Zacaro, [1]Alderico Rodrigues de Paula Junior and [2]Marcelo Lopes de Oliveira e Souza

[1]Research and Development Institute (IPD) Course of Biomedical Engineering, University of Vale do Paraiba (UNIVAP),Av. Shishima Hifumi, 2911, 12244-000, S. Jose dos Campos, SP, Brazil

[2]Course of Space Engineering and Technology (ETE),Division of Space Mechanics and Control (DMC),

National Institute for Space Research (INPE),Av. dos Astronautas, 1758, 12227-010, S. Jose dos Campos, SP, Brazil

ABSTRACT

Problem Statement

Significant effort goes into the development of biomedical software, which is integrated with computers/processors, sensors and electronic instrumentation devoted to a specific application. However, the scientific work on electronic instrumentation controlled by biomedical software has not emphasized software development, instead focusing mainly on electronics engineering. The development team is rarely composed of Software Engineering (SE) experts. Usually, a commercial automated tools environment is not used due to its high cost and complexity for researchers from other areas to understand.

APPROACH

This present study reports how the SE approach was applied to design and develop biomedical software, which is part of a Computerized Electronic

Instrumentation (CEI). This CEI comprises software and an electronic instrumentation based on a force sensor and electrogoniometer to monitor the hand exertion of computer user during typing task. The aim is to serve as a guideline for academic researchers who are not expert in software engineering methodology but usually develop their own software to run with their CEI. The specification of the requirements, presented as use case, includes the context diagram, the data flow diagram, the entity relationship diagram and test procedure. The Unified Modelling Language from the Enterprise Architect tool was used. The developed software and the electronic instrumentation were tested together.

Results

A sample of the interface screen shows how the outcomes could be plotted in an integrated manner. By comparing the values with other values obtained by manual calculations and with those provided by sensor manufacturer, the repeatability of test procedure validated the results. Reliable electronic instrumentation when working with unreliable software can become unreliable.

Conclusion

Applying software engineering methodology principles provided a simple and clear documentation that was helpful to establish the test procedures and the re-work.

INTRODUCTION

The scientific work in electronic instrumentation controlled by biomedical software has not emphasized the software development, but instead focused on electronics engineering, probably, because the development team is rarely composed of Software Engineering (SE) experts. However, in this study, software design and development was the target instead of electronic instrumentation development. This should help researchers in academic environments to better understand where the SE methodology approaches fit when they do electronic instrumentation that involves software development.

Biomedical Software and Mission Critical Applications

A lot of software is developed by people who are not experts in Software Engineering (SE) methodologies. A significant amount of software is developed by small and medium size software organizations, which do not have infrastructure and resources to implement a rigorous quality plan (Mishra and Mishra, 2009). Most cannot afford automated tools for SE due to their high cost and complexity. Therefore, design faults may occur as a result of imperfections from specification requirements. To avoid design faults, according to Troubitsyna (2010) while developing a system by refinement, developers start from an abstract specification. Stepwise refinement allows them to incorporate system requirements into the specification gradually and eventually arrive at system implementation, which is correct by construction. Design faults may cause errors and even failures in the system. Wrong results could be due to code implementation faults or faults in the specification.

Nevertheless, when a failure occurs in biomedical computerized electronic equipment, it is usually attributed to a protocol error, equipment failure, or human error during system operation. Rarely, is it attributed to a software fault that induced an equipment error. Early reliability models are based on reliability engineering, particularly hardware reliability (Shanmugapriya and Suresh, 2012). However, quality goals can primarily be achieved if the software architecture is evaluated with respect to its specific quality requirements at the early stage of software development (Shanmugapriya and Suresh, 2012). Redundancy is especially important in systems with critical missions. Software can have redundancy of processes, while hardware can have redundancy of electronic components such as sensors or input/output channels. A system is defined in the biomedical field as having a critical mission, if its failure could have life threatening consequences. Several authors have reported on such systems. Arpaia et al. (2012) proposed a method for home-care to predict a critical condition of a patient affected by a specific disease such as pulmonary disease.

Systems engineering can autonomously decide the next step for patient care based on the previous set. Kauffmann et al. (2011) offered approaches for verification and validation processes. The greater the risks, the more efforts must be used to reduce the probability of system failures. Depending on the mission of the system, its failure can result in lost of human life. For example, Garcia-Saez et al. (2009) discussed a solution for the problem of diabetes care based on architecture and implementation of a mobile personal assistant that supports personal and remote control strategies for insulin-dependent patients supervised by healthcare professionals through a telemedicine information system. The system notifies the patient with a message whenever a new therapy is prescribed. In this case, incorrect information would generate incorrect prescription, which could be ultimately lethal. Considering that an incident can occur when unprepared staff operates complex biomedical system, all biomedical systems should have access control for each available process into the system. One solution reported by Rezk et al. (2012) makes it possible to mine the dependency among user's data items from his transaction log and generates specific rules, which determine the context in which the user access the data items. According to them the challenge is building an efficient Database Intrusion Detection System, which can detect any malicious transaction with low false positive rate and high detection rate and integrating it with access control, in order to strength the database security. Two issues could minimize the problems reported above, one of them is the software engineering methodology; the other is the certification of software and electronic instrumentation.

Software Certification

Software certification is not usually taught in a university or a computer science course. Therefore, few software developers are prepared to develop software ready to be submitted to a certification process. Certification of medical software is not a recent approach; however, few researchers and governments have made any effort to do this. Certification allows software developers in the country to market safe products with better quality and, therefore, are more competitive. In addition, certification facilitates the entry of such software into international markets. For this purpose, the way that software packages are tested, delivered and operated should contain written document. The International Standard Organization (ISO)/12119:1994 is concerned about this. Additionally, the ISO 9001 and FDA medical device good manufacturing process regulations have roles that reduce the risk of low quality software production. Recent scientific study addressing certification of biomedical software that works with electronc instrumentation was not found. However, certification of systems containing software is increasingly important for governments, industry and consumers alike (Maibaum and Wassyng, 2008).

Guidelines from a certifier company are needed for biomedical electronic instrumentation that involves software development. Each certifier company has their own roles based on specific ISO standard. Due to the lack of roles based on ISO standard that could help researchers to conceive and develop their software for biomedical instrumentation, this computerized electronic instrumentation has not been submitted to a certification process yet. Nevertheless applying the SE methodology approach was an advance.

Software Engineering Methodologies

There are several approaches to SE and each lead to a specific methodology of SE. Structured Analysis of systems is a SE methodology used by several software developers. Nakanish et al. (2009) had the Data Flow Diagram (DFD) as their focus. The Agile is about rapid software design and development. The Scrum, based on agile methods, could be the best solution for project with rapidly changes and at first, it does not emphasize solid requirements as the structured analysis does. The ArchJava is a tool designed to allow programmers assemble components, connection and ports. An ArchJava component is a special object, able to be equipped with ports through which it can interact with other components. Ports are dedicated to support connections and service invocations. A port corresponds to the concept of interface found in COM port frequently used in communication between machines. Analysis and Object Oriented are different techniques to develop the computer system. The Object Oriented (OO) is the newest approach that focuses on capturing the objects in the scenario of the current system. A tool of modelling called UML (Unified modelling Language) serves for the use case and DFD definition that could be applied in both approaches of structured analysis and OO.

However, some researchers have developed biomedical software for CEI, but the focus of their manuscripts has been electronic instrumentation. Therefore, the reader or some new team member does not understand how the software was designed and developed. As a consequence, the system could become unrepeatable and not maintainable. Amato et al. (2009) developed a biomedical system using Java, relational MySQL DBMS, MATLAB and XML, but their SE design was not included. Pereira et al. (2009) developed instrumentation to apply and assess locomotor training. They constructed a system using electrogoniometers, load cell and software developed in LabVIEW® environment. The LabView is one software environment provided with its respective data acquisition card. Normally, the researchers generate documents applied to the electronic instrumentation only, while the software documentation is stored only in the mind of the developer. When the project moves out of laboratory and onto the market, the software should not be inserted into the project documentation as a "black box", but a systematic documentation of software is need to allow reproducibility of the project.

MATERIALS AND METHODS

Electronic Instrumentation

The computerized electronic instrumentation is an integrated solution that proposes to record typing frequency, fingertip force and wrist posture (flexion, extension, ulnar and radial deviation) in computer user to detect hand overexertion, during typing task, that could put the computer user at risk of acquiring muscle skeletal injury. Exactly how the electronic instrumentation was developed was addressed in our previous study: Machado and Villaverde (2011). Following is a brief description of electronic instrumentation. The electronic instrumentation is comprised a twoaxis capacitive sensor accelerometer±1.5g±90deg, with a 1Hz- 300Hz bandwidth response used as goniometer to register the wrist posture; and a Force Sensing Resistor (FSR), 4.4 N, with a response time of 5 μl, biased by dual ±9V and -5V power supplies used to register the applied force. An instrumented hand is shown in Fig. 1. This photograph was taken while a computer user was performing a typing test using the CEI. The accelerometer is in the box positioned on the subject's right hand. The force sensor resembles an almost transparent ribbon. An empty box was attached to the subject's left hand to simulate the condition of the test hand; simulated force sensors were also placed on the left hand.

After the hands were properly instrumented, both sensor circuits (force and accelerometer) were energized and connected to a data acquisition card (with 12 bits and 32 channels from Lynxtec Technology-Brazil) inserted into a personal computer. Figure 2 and 3 are the basic diagrams of circuit boards corresponding to the force sensors and the accelerometer, respectively. These diagrams show the output signal, which were acquired and processed by developed software

The V_{out} in Fig. 2 is the output signal in response to a pressure that changes the sensor resistance (R_1) to a reference resistance (R_2).

The accelerometer sensor has two reading exits, one for the X-axis and another for Y-axis; Y_{out} is the output voltage of the accelerometer, in Y direction; X_{out} is the output voltage of the accelerometer, in X direction. The accelerometer was used as a goniometer to measure wrist flexion/extension and ulnar/radial deviation. The developed software acquired and processed the output signal in order to provide the correspondent angle for wrist posture during typing task.

The software requires calibration tables to execute calculations using the acquired signal. Therefore, the software has a process responsible for sensors calibration. The calibration task was described in the Procedure for test and in the calibration outcomes section. With more calibration points, more precise results could be obtained, because the calculations were based on a linear interpolation method (Eq. 1):

$$X_2 = X_1 + (a - Y_{X1})\frac{(X'_1 - X_1)}{(Y_{X'1} - Y_{x1})}$$

(1)

where, variable "a" represents VOUT from Fig. 2 and V_{OUT}, X_{OUT} from Fig. 3; X_2 corresponds to the unknown fingertip force or wrist angle during typing task; while the interval value from the calibration table is X_1 with its respective tension YX_1 and X_1' with its respective tension $YX_{,1}$. The X_2 values were saved as temporary data for graphic generation by developed software.

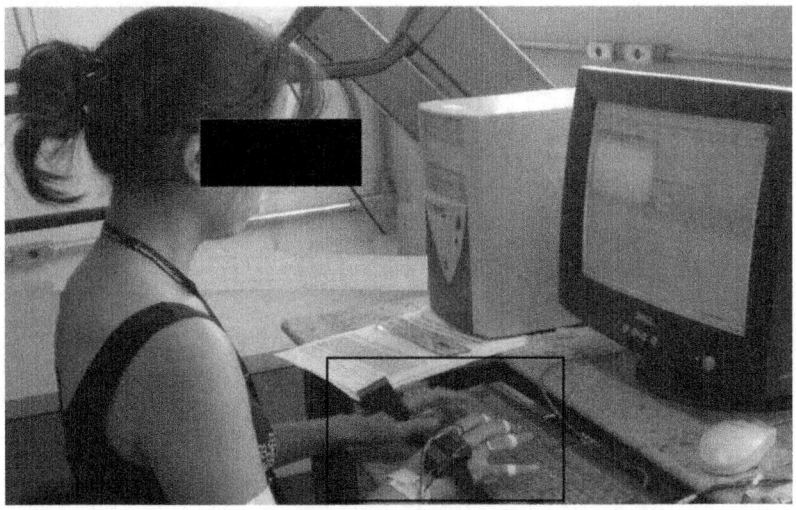

Figure 1: View of the instrumented hand showing the position of the sensors.

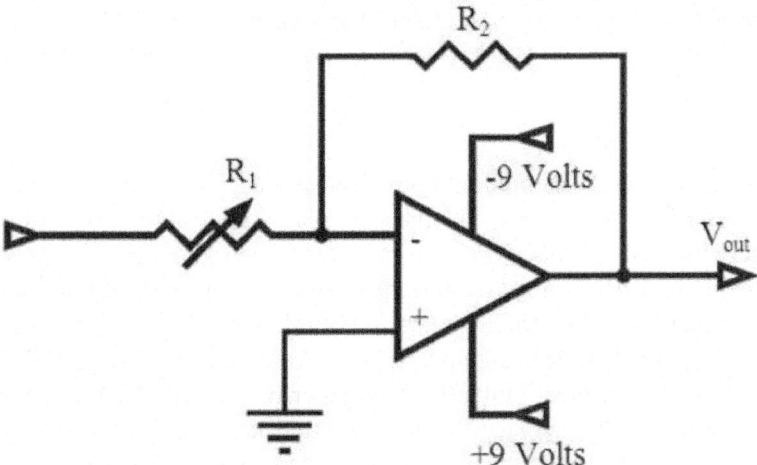

Fig. 2: Basic diagram for force sensor. Legend: R_1 = sensor resistance; R_2 = Reference resistance; V_{out} = output signal

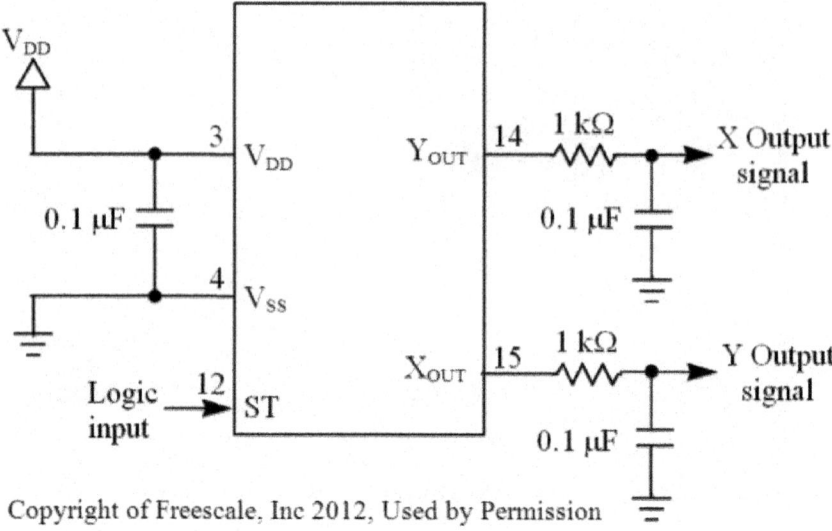

Copyright of Freescale, Inc 2012, Used by Permission

Figure 3: Connection Diagram for Accelerometer. LEGEND: VDD=the power supply input; VSS=the power supply ground; ST=logic input pin used to initiate self test; X_{out}=output voltage of the accelerometer; Y_{out} =output voltage of the accelerometer.

Software Design

The structured analysis methodology was applied. The software requirements were described by use case. The use case diagram is commonly applied to the Object Oriented approach. For the structured analysis approach, the software requirements are described by narrative use case as a series of numbered steps and complemented by context diagram. The relational database manager system Firebird®; the Delphi® language; Unified Modelling Language (UML) from the Enterprise Architect tool; a PC Computer with Microsoft Windows, a data acquisition card by Lynxtec Technology; and the electronic instrumentation previous described were used. The first step of this methodology was the narrative use case as a series of numbered step of software requirements, present in the results section. The second step is the context diagram (Fig. 4) design that can be helpful in understanding the context that the system will be part of.

The next step was the Entity and Relationship Diagram (ERD) designing (Fig. 5-6). This ERD is a relational model to show a brief database composition and how the data are inter-correlated. Each entity was represented by a rectangle; while each data was defined as an attribute of its respective entity, known as table column. For each entity created, its attributes were defined and listed on the right side of Fig. 5 and represented by an ellipse in Fig. 6. The designed ERD had an indicator of its cardinality. The database modeling was performed using UML Data Modeling Profile. This tool maps the database concepts of tables and relationships onto the UML concepts of classes and association by using the stereotype.

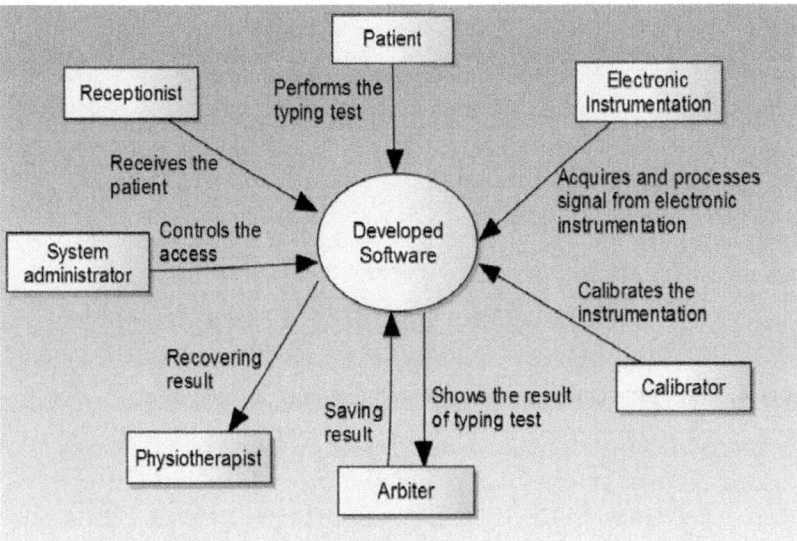

Figure 4: Context diagram of the system.

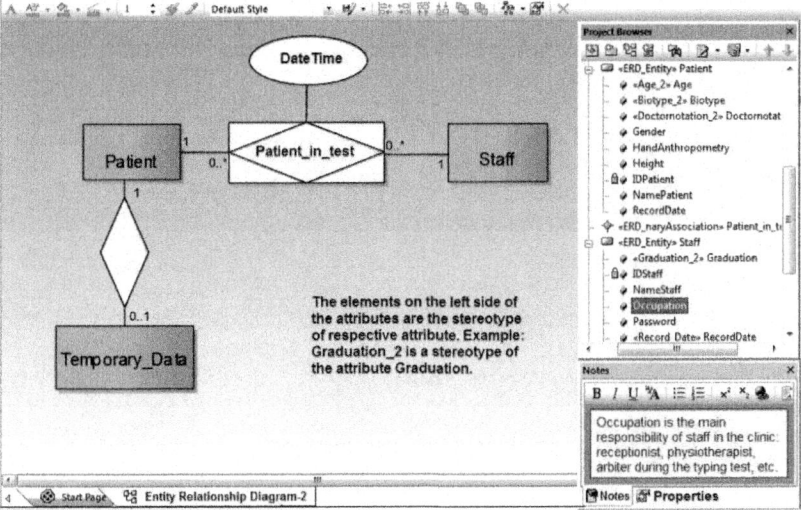

Figure 5: The used tool case showing the starting of the entity and relationship diagram definition and its respective attributes.

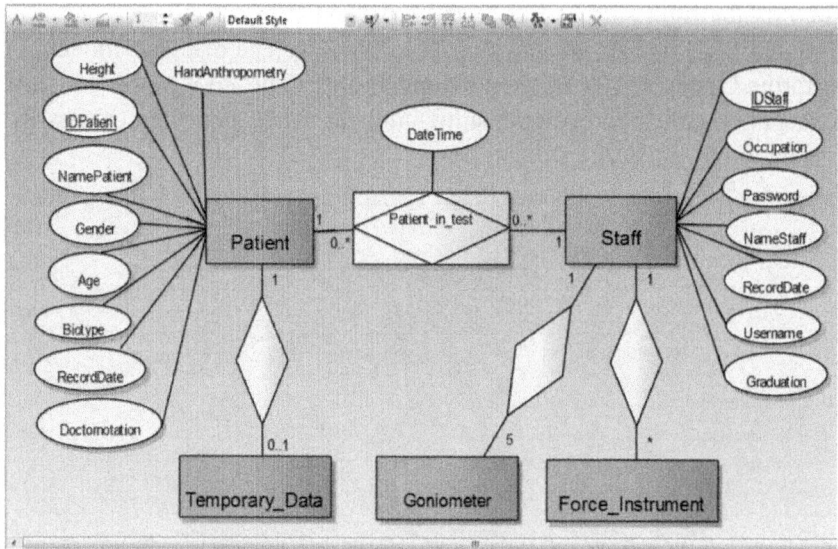

Figure 6: Part of the entity and relationship diagram of the system.

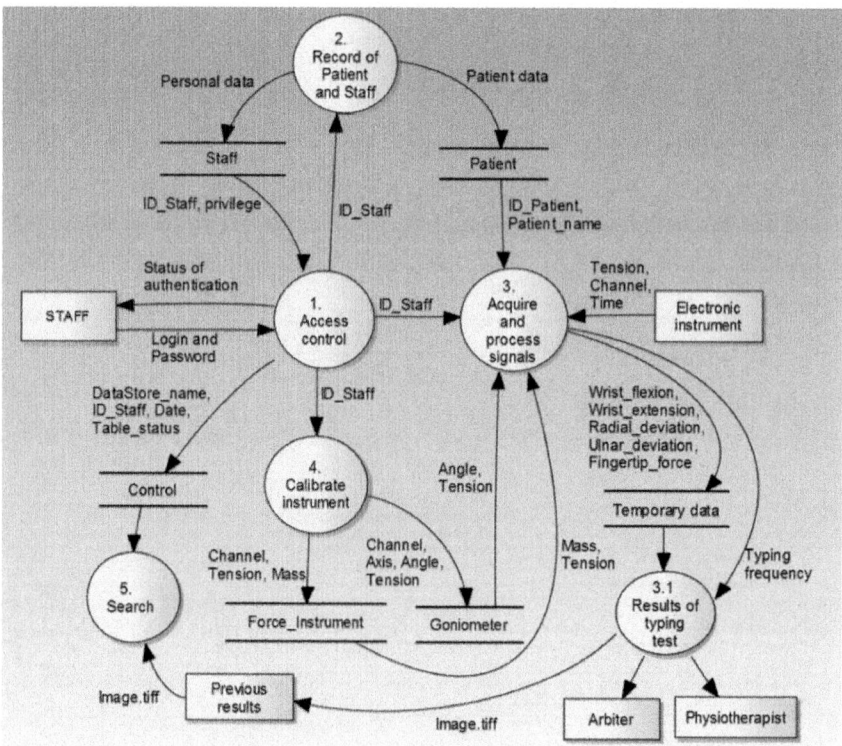

Figure 7: Data flow diagram of the system-top level.

The fourth step of this SE methodology was to design the Data Flow Diagram (DFD), shown in Fig. 7, composed by process, data storages and external entities. This DFD helps visualize how the system should operate and how the system would be implemented. The used tool case provided the notation used for DFD design, where the ellipse represent a process in which data are used or generated; the rectangle represents an external source; the two parallel lines represents the stored data; and the arrow indicates the direction of data, i.e., how data flows through the system.

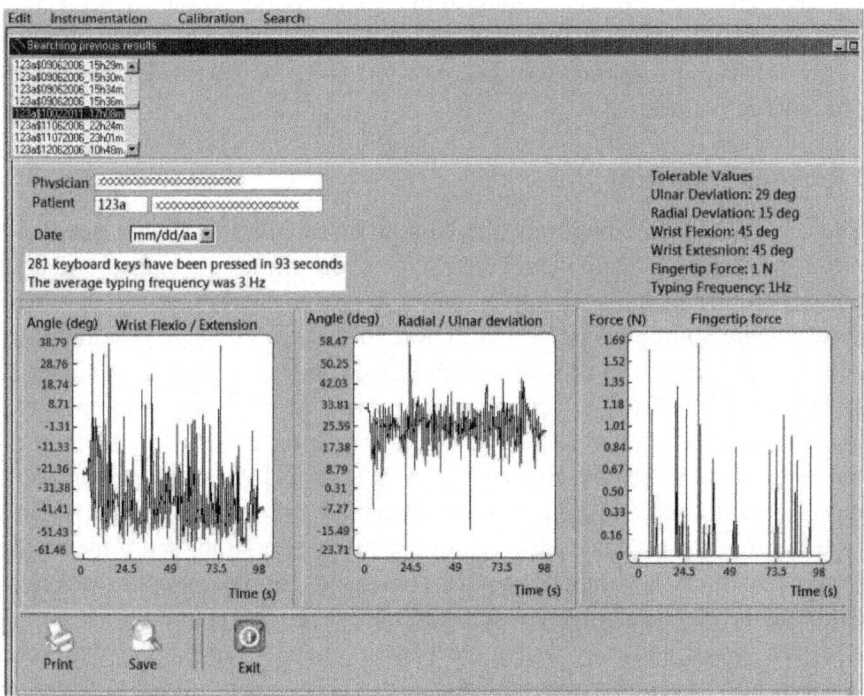

Figure 8: How the outcomes were exhibited in an integrated manner. LEGEND: Deg=degree that refers to the threshold for wrist posture; N=Newton that refers to the fingertip force applied during typing; Hz =the average of keyswitch pressed in one second.

The fifth step was the software implementation by code creation using the Delphi® language.

Code Implementation

The driver of the acquisition card had its source code adapted to run as part of our software, to provide facilities to choose the channel and sample frequency, as well as to acquire and process data. Process 3 from Fig. 7 was programmed considering the pins where the sensors were connected to the data acquisition card. These signals were processed and stored in a Table as "Temporary-data" according to Fig. 5-7. The access control process from Fig. 7 was the first to be implemented enabling and disenabling access to the other options of the system. Each process was implemented and tested in isolation. Finally, the integrated system was obtained when the electronic instrumentation was connected to the computer. Then, process 3 was tested to evaluate its capability to recognize the electronic instrumentation. Each channel from the data acquisition card was tested to assure the correct reading from each connected sensor. The electronic instrumentation and software were tested together to evaluate the capability of the software to acquire and process data. The outcomes were compared to previous set values.

Procedures for Test

The test team was composed of a computer scientist (software developer), electronic engineer and physiotherapist. To check the access control and assure that all requirements were implemented, a person, who was unauthorized to access some source of the system, checked the security/safe condition. Two people were the system testers. Each logged in the system as administrator and all permission to the administrator profile was tested. This same person logged out as administrator and subsequently, logged in as receptionist; then the type of permissions granted her was tested. This same person logged out as receptionist and logged in as arbiter so she could setup the software configuration to recognize the electronic instrumentation, setup the channel from where the data would be acquire; choose the sample frequency before use; calibrate the electronic instrumentation; start a simulation of a test by pressing all force sensor and moving the electrogoniometer, stop the simulation; and observe how the outcomes were exhibited. The second tester repeated this sequence.

To validate the result provided by software and the electronic instrumentation working together, one force sensor was tested by using standard masses, when 23 calibrated points were obtained starting from 20-430 grams. For this purpose, the force sensor was connected to the data acquisition card inserted into the computer. The circuit with the force sensor was power supplied by an electronic device with a dual voltage of ±9V. The sensor, with a sensitive diameter, was placed on a flat area. The masses were placed on the sensitive perimeter of the sensor and the response was shown in tension through the software. When a pressure was exerted on the sensitive area, a reduction in its resistance occurred and, consequently, a voltage signal was released. To validate its reliability, the same force sensor was tested three times. The outcomes are presented in Table 1. The mean and standard deviation was calculated between the three trials to observe the error rate in order to estimate the reliability of the software in providing accurate results from electronic instrumentation in response to the fingertip force applied during typing task.

Table 1: The calibration table obtained during three trials with one force sensor.

Masses (gram)	Trial-1 (volt) (volt)	Trial-2 (volt)	Trial-3 (volt)	Mean (volt)	SD
20	0.23	0.21	0.28	0.24	0.04
50	0.61	0.58	0.45	0.55	0.09
70	0.74	0.80	0.70	0.75	0.05
100	1.10	1.00	1.10	1.07	0.06
120	1.20	1.40	1.40	1.33	0.12
130	1.30	1.40	1.60	1.43	0.15
150	1.40	1.50	1.70	1.53	0.15
170	1.70	1.70	2.00	1.80	0.17
180	1.90	1.90	2.10	1.97	0.12
200	2.40	2.50	2.50	2.47	0.06
220	2.50	2.50	2.60	2.53	0.06
230	2.60	2.40	2.60	2.53	0.12
250	2.90	2.80	2.80	2.83	0.06
270	3.10	2.90	3.18	3.06	0.14
280	3.10	3.00	3.42	3.17	0.22
300	3.30	3.30	3.60	3.40	0.17
320	3.70	3.70	3.80	3.73	0.06
350	4.20	4.20	4.00	4.13	0.12
360	4.22	4.21	4.25	4.23	0.02
370	4.60	4.50	4.30	4.47	0.15
380	4.60	4.50	4.50	4.53	0.06
400	4.70	4.90	4.60	4.73	0.15
430	4.92	4.98	4.85	4.92	0.07

To obtain the calibration points from the electrogoniometer, it was connected to the computer, the software responsible for calibration task was set up and the electrogoniometer was positioned at 0° and ±90 degree for x-axis and y-axis. Then the outcomes were transferred to the calibration data storage labelled as Goniometer in Fig. 5-6. Five calibration points were recorded. To check the reliability of this present software in calculating the typing frequency, no sensors were used; one subject typed a text using the computer keyboard for 90 s. Then, the system showed the typing frequency expressed in Hz, which was compared with manual calculation.

To check the accuracy of software in acquiring and processing signals, one computer user sat in an upright position with arm held at a flexion of 90 degree to the body. The forearm was in the resting position over a flat area, leaving the left hand free to move by flexing the wrist. The electrogoniometer was fixed onto the volunteer's hand, at the metacarpal junction of the middle finger and then this was connected to the computer through the data acquisition card. The software was configured at a sampling rate of 5 Hz. At the same time, a manual goniometer was placed on the side of the subject's hand, in the carpal area, with its movable spindle positioned at the edge of the little finger following the median line of the ulna. The subject did some wrist flexion movements, as requested by the arbiter. The manual goniometer followed the wrist flexion. The result from electrogoniometer was compared with the result from manual goniometer. This test procedure is the same reported by Machado and Villaverde (2011). Later, the test procedure was set up with one computer user. For this purpose, the electronic instrumentation was positioned on the subject's right hand at the metacarpal junction of the middle finger (Fig. 1) to register wrist posture. With the subject's hand in a prone position, a flex force sensor was fixed with a ribbon sensor on the palmar aspect of fingers 2 through 5 in the medial phalanx region with the sensitive area of the sensor on the fingertip pulp to register the applied force. Fig. 1 shows an instrumented hand. After the hands were properly instrumented, both sensor circuits (force and accelerometer) were energized and connected to a data acquisition card inserted into a personal computer. The software was setup to enable the channels from data acquisition card and the sample frequency was established at 2000 Hz. The subject started typing a previously selected text and the software started to acquire and process data, ending as soon as the arbiter pressed the "Stop" button available on the interface screen.

The outcome was shown in an integrated manner on a computer screen, which is illustrated in Fig. 8. These included information about the typing frequency, wrist flexion and extension, radial and lunar deviation and fingertip force.

The outcomes were saved as image.tiff, as predicted in Fig. 7. The software offers a resource for recovering this image.

RESULTS

This methodology was applied in a simple form, which is understandable by anyone who intends to develop biomedical software for electronic instrumentation. The narrative use case, the context diagram, ERD, the DFD, the result of the calibration task and a sample of the outcomes provided by CEI during typing task in attendance with item 21 from use case are shown below.

Narrative use Case of the Software Requirements

The system administrator creates the staff records for receptionist, arbiter, calibrator and physiotherapist to provide user authentication:

- One patient first arrives to do a test
- The system requires the receptionist authentication
- The system verifies the user authentication profile in order to enable or disable resources
- The system enables the receptionist to create the patient record
- The receptionist receives the patient by providing a patient record at his/her first time. The receptionist then logs out from the system
- The instrument calibrator logs in the system to obtain user authentication
- The system verifies the user profile to enable the screen interface for calibration task
- The instrument is connected to the computer
- The calibrator calibrates the force sensors and the electrogoniometer to actualize the database with the new calibration data and logs out of the system
- The system registers any changes made in the database
- The arbiter/physiotherapist logs into the system
- The record of respective patient who arrived to do the typing test is selected
- The subject sits upright in a chair with arms and knees flexed at 90 degree. The keyboard is approximately 10 cm from the edge of the table. The subject starts typing a previously selected text for 90s. The developed software obtains the typing frequency
- The typing frequency is saved by system
- The patient has the hand instrumented with force sensor and electrogoniometer attached
- The arbiter decides the duration of the test, selects the channel for signal acquisition an chooses the sample frequency. Then the patient starts the typing task
- The software starts the control of the channel/sensor, registers the signal in volts, records the reading time and transforms the values obtained in volts into the correct measurement units (degree, Newton and Hz).
- The result of calculation is retained in a temporary Table.

- The arbiter administers the use of the instrument, tracks completion of the test and stops it by pressing a button available on user interface screen
- The system shows the outcomes on a screen interface. Data on flexion and extension of the wrist, radial and ulnar deviation, typing frequency and fingertip force during the typing are exhibited
- The arbiter saves or prints the result as an image
- The arbiter/physiotherapist and patient analyse the result together
- The system provides limits for wrist flexion and extension, lunar and radial deviation, fingertip force and typing frequency above which the patient is in risk of acquiring muscle skeletal disorder
- The physiotherapist researches previous result of the respective patient
- The patient receives guidance about extreme values observed
- The physiotherapist fills the electronic record with information related to the diagnostic conclusion
- The physiotherapist/ arbiter logs out from the system
- The system saves information about user name, date and database structure in which the respective user has just modified. The system allows the administrator to audit the database to assure data safety

Context Diagram

The Context Diagram (Fig. 4) has just one process element representing the system being modelled, showing its relationship to external systems, person, equipment, or something that is necessary during software use.

The Context Diagram is Composed of Previous results-were saved in one folder into the hard disk of the computer where the screen of the outcomes was saved as image.tiff. The purpose of this is to track patient's improvement concerning hand overexertion. Staff-people with authorization to operate the system and therefore register patients, be responsible for CEI use, search for previous results, analyse the outcomes. In summary, staff is a person who is not patient, but is involved in the typing test environment:

- Arbiter-health professional who is responsible for instrumentation use and analyses of the patient's test outcomes
- Electronic Instrumentation-the CEI composed by force sensor, electrogoniometer, data acquisition card and power supplier
- Patient-the computer users who will use the CEI solution to detect his/ her vulnerabilities in acquiring muscle skeletal disorder

- System administrator-person responsible for database system and software performance and security
- Computer-used to run the developed software including the goniometer, force sensors and the data acquisition card. The typing task test was also performed on this computer

Entity and Relationship Diagram

There was no intention to write a tutorial; therefore, Fig. 5-6 is a brief demonstration of designed ERD. Consequentially, not all created entities were represented in Fig. 6. In Fig. 5, part of the tool case is shown to provide an idea how the entity and its respective attributes were created. In this case, the entity Patient and Staff are shown. Figure 5 (b) is part of the Entity and Relationship Diagram of the system in a more advanced stage of creation than Fig. 5. The diagrams in both Fig. 5-6 are not a flowchart. Therefore, the losangle does not represent decision made: this represents the relationship between the two entities. The relationships illustrate how two entities share information in the database structure. The relationship between entities Patient and Staff was the primary key from Patient and Staff, plus a new attribute labelled as Date Time that denotes the date of patient attendance. For each data storage in this ERD, there is an equivalent data storage in the DFD. The underlined attribute (Fig. 6) represents a primary key. These keys were used for relationships between the data storages and warrant uniqueness. The data storage named "Patient" has information about each patient. Whereas, the data storage named "Temporary_Data" contains processed signals obtained during typing task. The data storages named "Force_Instrument" and "Goniometer" were used for data storage from the instrument calibration; staff with calibrator profile is responsible for these table's contents. This ERD diagram shows the attributes distributed among entities that represent the data storages, illustrated with rectangles. In reality, each data storage in ERD had an equivalent data storage in DFD; and each data storage generates one table in the Firefird® database. Two entities are shown on the left side of Fig. 6 to demonstrate how the attributes were designed by using an ellipse and line connector to the respective entity. The attributes of each respective entity are listed on the right side of Fig. 5. The relationship named of "Patient_in_test" from Fig. 6 enables the system to provide information about patient's attributes and consultation date, as well as the attributes of involved staff. Additionally, when the staff has the health care profile defined by the occupation attribute, they have free access to the Temporary-Data. There are many notation styles that express cardinality. The tool used for this ERD designing provides the used notation, where:

- 0..1 - zero or up to 1 instances, but no more than 1
- 0..* - zero, one or many instances
- 1..* - 1, or more than 1 instances
- - exactly 1 instances

Data Flow Diagram

To show the solution of the SE, the Data Flow Diagram (DFD) was designed with several levels. However, this study only shows the DFD in the top level of abstraction, i.e., level-1 composed by process, data storages and external entities. The developed software was divided in many parts. Each part was responsible for one or more tasks from the use case. The DFD diagram in Fig. 7 shows each part of software defined as numbered process (1, 2, 3, 4 and 5). Process-1 was created for user authentication and access management to each process in the system, guaranteeing set restrictions for system use. The user's profile determined which processes became available. Inserting a new patient record into the database was a task for staff, who are profile receivers. This key process works as a connector between other processes where security services were performed to guarantee confidentiality, integrity, access control and non-repudiation. Process-2 is for registering patients or authorized staff to operate the instrument.

Process-3 controls the sampling frequencies and the number of samples in each channel. In addition, the calibration values obtained from process-4 are used to calculate values from hand exertion. Process-4 helps to calibrate the instruments used for angle and force measurements. The respective calibration values were then stored in data storage Goniometer and Force_Instrument. Process-5 performs support tasks such as searching previous patient test files, which were saved in the computer in an image format and controlling the alterations made in the database, making it possible to gather information about who changed the data.

Data storage-Patient contains register information about patients; and data storage-Staff is to record the staff involved in the system operation. Temporary_Data is used to save temporary data about the acquired signal of hand exertion. Changes in the database, from any part of the system, are automatically recorded in data storage labelled as Control.

The calibrations Outcomes Performed via Software

The measurement of wrist posture with a goniometer does not require decimal place precision (i.e., integer number). Oscillations or noise (drift problems) were observed in the fifth decimal places from the measured values. The used electrogoniometer had an offset in X and Y-axis of 3 and 6°C, respectively, when checking the accuracy of software in acquiring and processing signals, which was considered during software development. The test with software and instrumentation for fingertip force measurement when repeated three times showed a mean standard deviation of 0.10 volts and the maximum was 0.18 volts (Table 1). These are compatible with the expected error rate (±5%) reported in the datasheet from the manufacturer of the force sensor. For the calibration task, calibration table became empty automatically, because all previous points must be recalibrated. The history about who did what was recorded in the data storage for access control. A staff with system administrator profile had access to all information, especially to the access control, defined as Control in Fig. 7.

A sample of outcomes provided by software and electronic instrumentation together: The model of exhibition of integrated outcomes obtained from electronic instrumentation is exhibited in Fig. 8. Figure 8 shows the outcomes: patient/subject code, patient/subject name; arbiter name; typing speed; typing frequency; flexion/extension of the wrist; lunar/radial deviation; fingertip force; tolerable values to be reached; and test date. This information helps the health professional to diagnose the computer user who is inclined to acquire the Work Related Upper Extremity Disorder (WRUED) due to inadequate hand exertion. Improvements were done to this software in response to Machado and Villaverde, (2011), which did not successfully measure radial and lunar deviation. In the upper left corner of Fig. 8, a sample shows how the saved outcomes were recovered. The saved file name was composed of subject code plus date and time of test, separated by special characters. The tolerable values shown in Fig. 8 were set by Machado and Villaverde (2011). The subject and arbiter analyse the outcomes regarding these tolerable values in order to make decisions for extreme hand exertion to avoid the risk of inciting muscle-skeletal illnesses. Only one button needs to be pressed to save the test results, with the file name and location automatically chosen by the developed software.

DISCUSSION

Applied software engineering methodology: Any SE methodology approach could be an advance. The developers would choose the SE methodology and one appropriated tool case. Consequently, comprehension about what/how the software does provides information useful for software maintenance and validation method. Applying software engineering methodology, the requirements were easily traced after the software implementation. Moreover, perceived errors during the software test were quickly located in the respective source code and resolved. According to Constantine and Lockwood 2012, one advantage of narrative use case is that a series of numbered step is immediately apparent, because the separation into distinct steps makes it easier to skim the use case for an overview and the general nature of the interaction." Functional similarity was not found in other software. Nakanish et al. (2009) presented a DFD, but their ERD diagram was not found, nor was a sample of their software requirement shown. Therefore, it was impossible to apply the same methodology of SE as they did. Nevertheless, their explanation about the DFD designed was fully comprehendible. Amato et al. (2009) seem to have used a systematic procedure. Respective documentation probably was done; however, their SE design was not included in their manuscript. On the other hand, the present study disclosed our software engineering design to encourage research of electronic instrumentation to make software documentation. No specific tool case is mandatory. There are several tool cases, some is totally free while others are temporary free provided as a demo-version. There are also the open source tools devoted to the designing and development of software that use open source platform such as Linux. There are several commercial software to acquire and process signal. However, they do not allow changes to the source code in order to obtain personal software. Therefore, the requirements presented in the case use, would become limited to the scope of the software. This was one more reason to decide to develop the present software. The software for the signal acquisition was modified and embedded into this present system to be part of them. This software solution generated low production cost and anything can be changed at any time.

Contrary to this, Pereira et al. (2009) developed instrumentation and built three software programs using LabVIEW® environment. According to them, this tool provided agility to building the system but was certainly more expensive to reproduce. No documentation about software design was reported. However, they intend to develop their software in the conventional manner for lower production costs. However, if the same researcher is not available anymore, due to the lack of software documentation, more time would be demanded during software rebuilding. Hence, this present study can hopefully make the researchers realize the advantages in applying some SE methodology. There are modern methodology for software development that could be used to develop biomedical applications (Object Oriented-OO, Artificial Inteligency-AI, Agile methods, SCRUM). However, when a methodology is new, the probability to find a partner or professional to work on the team could be reduced. In addition, depending on the application, one methodology fits better than another. The OO approach is used less than structured analysis among veteran software developers, while AI is an approach specific for application that supports decision-making. The Agile methodology emphasizes communication among developers and users, rapid changing of information and adaptability to the change. The Scrum is an interactive and incremental agile software development method. The Aegis methods are for rapid software development, after that, the details of requirements are implemented gradually based on customer opinion, making it difficult to predict anything before the end. According to Hajjdiab et al. (2012), the lack of knowledge transfer between team members contributed for unsuccessful adoption of Scrum method in the United Arab Emirates. This may be due to the Scrum being a rarely used method. Conversely, for about three decades, structured analysis has been used; consequently, the possibility of find experts is larger. Structured analysis is based on process and relational database, also a consistent requirements definition is expected when changing of the requirements is not expected during implementation. According to the chosen SE methodology, the specific programming language and tool case was chosen. This present work was based on structured analysis methodology. The Enterprise Architect tool case was chosen to design this present software because its free download was possible. However, for this computerized electronic instrumentation to be in marketing, the purchasing of commercial version of used tool case is needed.

When software is for health care, it was important to use a solid and safe database, as an example Oracle, Microsoft SQL, Firebird® , My SQL, PostgreSQL. Some of DBMS above are free while others must be purchased. Some specific training is necessary to work with any of these technologies. We used the Firebird ® , because it was totally free in this version. The use of a solid and safe database does not warrant data protect. Some specific training is necessary to work with any of these technologies. Few works have been done about protecting data for confidentiality, integrity and access control to sensitive information as those discussed by Rezk et al. (2012). In their purpose, when user submits a query to execute, the system checks if he has authority to access data items in the query, the context in which the user accesses the data items is checked to determine if he follows his normal behavior or not. In our case, the access control protects data by verifying the staff authentication profile in order to enable or disable resources. Also, this software controls the alterations made in the database, making it possible to gather information about who changed the data, indicating the user ID, data and time and the name of changed table.

Calibration Outcomes Performed via Software

The sensor response capability was a linear curve; however, its predicted error rate could propagate among the acquired and processed signal, which could contribute to the standard deviations in the Table 1. The three trials present a mean standard deviation of 0.07 grams, taking into consideration the error rate of±5% in full scale predicted by the sensor manufacturer. Therefore, considering the full scale during the calibration section represented by 430 grams, producing mean tension of 4.92 volts (Table 1), the error rate of 5% was calculated. The outcomes showed that a standard deviation of 0.24 volts could be expected during calibration section. Consequentially, the reliability of software in providing accurate results for fingertip force measurement can be taken into consideration, because the maximum standard deviation observed among the three trials during calibration section was 0.14 volts. The electrogoniometer working with developed software was previously tested and reported in Machado and Villaverde (2011). The result had an average deviation of 1.24 degrees between the electrogoniometer and the manual goniometer. This was considered satisfactory if compared with the nonlinearity of ±1% of its out signal predicted by sensor manufacturer.

Integrated use of Software and Electronic Instrumentation

The general mechanism of error detection is to intercept outputs produced by a system (or a component) and to check whether those outputs conform to the specification (Troubitsyna, 2010). Manual calculations were used to validate the results relating to fingertip force, wrist flexion and extension, ulnar and radial deviation and typing frequency obtained by the software. Similar validation had already been efficiency applied by Kauffmann et al. (2011). It is extremely difficult to validate software that does not have written requirements. Because the software and electronic instrumentation work together, a problem in one can make both become unreliable. We experienced this when the unsuccessful measurement of radial and ulnar deviations were attributed to the accelerometer technology. Therefore, another accelerometer was provided, but the problem remained, because this problem, attributed to an equipment failure, was actually due to a software implementation problem. A consultation with an expert in measuring with an accelerometer concluded that implementation code for wrist flexion and extension measurement is easier than radial and ulnar deviation, which requires signal composition during code/ software implementation to reach accurate result. His suggestion was applied. This occurrence showed that human error during requirements specification or during design process can cause wrong results. This kind of error is not so easy to detect during the testing phase, because everything works normally and no error appears.

A challenge during developing of biomedical software is the requirements. For any software developer, the problem domain is very complex. The performance of our software working with electronic instrumentation reached the expectation of the physiotherapist, who used the CEI as a volunteer and arbiter. The use of this system by a physiotherapist was an important phase of the test. Upon her suggestion, changes in software occurred to use the correct medical vocabulary on the interface screen. A friendly software interface encourages the users to carry out instrumentation manipulation. Further implementation would adopt specific database for image to prevent forced access by an unauthorized user to the test result provided in the Fig. 8 and stored as image. Another future improvement is to have the tolerable values presented as lines in the graphical output from Fig. 8, allowing the subject to train to within those limits and adjust his/her movements immediately." This software guaranteed that the data generated and medical diagnoses remain private. According to Shanmugapriya and Suresh (1012), a software system's reliability is defined as the probability of the software operating without failure for a specified period of time in a specified environment. One example is the system reported by Garcia-Saez et al. (2009) that supports remote control strategies for insulin-dependent patients. In this case, unreliable system would let to incorrect prescription that could be ultimately lethal. Because this present software was devoted to an application that does not fit into a critical mission, in that its failure will not result in loss of human life, it was not necessary to estimate the order of magnitude for probabilities of failure per hour. Computerized electronic instrumentation should be certified in order to put it on the market. If an organization isn't worried about safety, it must consider the consequences of using mission-critical software that isn't certified or qualified as fit for purposes (Maibaum and Wassyng, 2008). For an electronic instrumentation to be marketed in Brazil, it must be submitted to the Brazilian National Health Surveillance Agency (Agência Nacional de Vigilancia Sanitária-ANVISA) to obtain registration/authorization. This is because some instrumentation could cause damage or harm to the patient. In Brazil, the certification of biomedical electronic instrumentation can be obtained from a certifier company and INMETRO (Instituto Nacional de Metrologia). INMETRO certifies the accuracy and capability of the electronic instrumentation to accomplish its purpose. In addition, INMETRO evaluates if the risk of accident exists. Normally, Certifier Companies follow some ISO standard, which are not easy to understand and accomplish to make the system ready to be submitted for a certification process. Our software does not fulfil ISO standard, due to the difficulty in both adequate software and hardware. Mishra and Mishra (2009) reported a simplified software inspection process in compliance with international standards for software quality assurance. However, this is not enough to reach software certification. The certification of medical software is a difficult due to limited experience of the software developer or the difficulties in accessing and applying the standards. This is one challenge to be reached by the authors from this present study. This would be useful if academic researchers would publish their experience while submitting their CEI to a certification process. This present study describes how the software engineering methodology was applied to design and develop software to acquire and process sign from a computerized electronic instrumentation. In both case, the authors intend to offer their experience as a guideline to another researchers. The way that this study was reported could serve as a guideline for academic researchers who are not experts in software engineering methodology but usually develop their own software to run with their prototype of electronic instrumentation. However, this study cannot provide an example of how to obtain a software certification.

CONCLUSION

Development of biomedical software to work with electronic instrumentation needs to pay attention to both performance and system accuracy. This software was developed to acquire and process signals from a specific instrumentation for measuring hand exertion to help the computer user detect their overexertion. The aim was to demonstrate how to apply the structured analysis in biomedical software to acquire and process signals when the researcher is not expert in SE methodologies. Usually, when the developer leaves the team, no one is able to maintain the software due to a lack of documentation. One capacity of SE is to produce software that is maintainable and reusable due to its clear documentation. Due to the lack of guidelines on how to obtain certification of biomedical software that works as part of computerized electronic instrumentation, this present system was not submitted to a certification process. Despite this, the systematic documentation provided while the SE methodology was applied was an advance.

ACKNOWLEDGEMENT

This study was supported by Wagner Lima dos Santos (computer science) from WLSantos and Cia Ltda. Michelle Fernanda de Lima (Physiotherapist); Dr. Ricardo Toshiyutki Irita (Electronic engineer) from the Instituto Nacional de Pesquisas Espaciais-Brazil; and Alene AlderRangel from Univap (English support).

REFERENCES

1.   Amato, F., M. Cannataro, C. Cosentino, A. Garozzo and N. Lombardo et al., 2009. Early detection of voice diseases via a web-based system, Biomed. Signal Process Control, 4: 206-211. DOI: 10.1016/j.bspc.2009.01.005

2.   Arpaia, P., C. Manna, G. Montenero and G. D'Addio, 2012. In-Time Prognosis based on swarm intelligence for home-care monitoring: A case study on pulmonary disease. IEEE Sensors J., 12: 692-698. DOI: 10.1109/JSEN.2011.2158305

3.   Constantine, L.L. and L.A.D. Lockwood, 2012. Structure and style in use cases for user interface design.

4.   Garcia-Saez, G., M.E. Hernando, I. Martínez-Sarriegui, M. Rigla and V. Torralba et al., 2009. Architecture of a wireless personal assistant for telemedical diabetes care. Int. J. Med. Inform., 78: 391-403. DOI: 10.1016/j.ijmedinf.2008.12.003

5.   Hajjdiab, H., S. Al-Taleb and A. Jauhar, 2012. An industrial case study for scrum adoption. J. Software, 7: 237-242. DOI: 10.4304/jsw.7.1.237-242

6.   Kauffmann, C., A. Tang, A. Dugas, É. Therasse and V. Olivab et al., 2011. Clinical validation of a software for quantitative follow-up of abdominal aortic aneurysm maximal diameter and growth by CT angiography. Eur. J. Radiol., 77: 502-508. DOI: 10.1016/j.ejrad.2009.07.027

7.   Machado, G.A.L. and A.J.B. Villaverde, 2011. Design of an electronic instrumentation for measuring repetitive hand movements during computer use to help prevent work related upper extremity disorder. Int. J. Ergon., 41: 1-9. DOI : 10.1016/j.ergon.2010.11.003

8.   Maibaum, T. and A. Wassyng, 2008. A productfocused approach to software certification. IEEE Software Technol., 41: 91-93. DOI: 10.1109/MC.2008.37

9.   Mishra, D. and A. Mishra, 2009. Simplified software inspection process in compliance with international standards. Comput. Stand. Inter., 31: 763-771. DOI: 10.1016/j.csi.2008.09.018

10. Nakanish, T., Y. Tsuchiya, T. Sakamoto and A. Fukuda, 2009. Structured analysis for software product lines. Proceedings of the 13th International Symposium on Consumer Electronics, May, 25-28, IEEE Xplore Press, Kyoto, pp: 915-919. DOI: 10.1109/ISCE.2009.5157027

11. Pereira, E., E.F. Manffra, J.A.P. Setti, C.M.R. Dutra and L.R. Aguiar, 2009. Development of instrumentation for application and assessment of locomotor training with partial body weight support. Brazilian J. Biomed. Eng., 25: 185-197.

12. Rezk, A., H.A Ali and S.I Barakat, 2012. Database security protection based on a new mechanism. Int. J. Comput. Appli., 49: 31-38. DOI: 10.5120/7879-1188

13. Shanmugapriya, P. and R.M. Suresh, 2012. Software architecture evaluation methods-A survey. Int. J. Comput. Appli., 49: 19-26.

14. Troubitsyna, E., 2010. Developing fault tolerant distributed systems by refinement. Proceedings of the 5th International Conference on Software Engineering Advances Software Engineering Advances ICSEA, Aug. 22-27, IEEE Xplore Press, Nice, pp: 178-183. DOI: 10.1109/ICSEA.2010.34

Chapter 7

INSTRUMENTATION FOR FERROMAGNETIC RESONANCE SPECTROMETER

Chi-Kuen Lo

Department of Physics, National Taiwan Normal University, Taipei, Taiwan

INTRODUCTION

Even FMR is an antique technique, it is still regarded as a powerful probe for one of the modern sciences, the spintronics. Since materials used for spintronics are either ferromagnetic or spin correlated, and FMR is not only employed to study their magneto static behaviors, for instances, anisotropies [1,2], exchange coupling [3,4,5,6], but also the spin dynamics; such as the damping constant [7,8,9], g factor [8,9], spin relaxation [9], etc. In this chapter a brief description about the key components and techniques of FMR will be given. For those who have already owned a commercial FMR spectrometer could find very helpful and detail information of their system from the instruction and operation manuals. The purpose of this text is for the one who want to understand a little more detail about commercial system, and for researchers who want to build their own spectrometers based on vector network analyzer (VNA) would gain useful information as well.

FMR spectrometer is a tool to record electromagnetic (EM) wave absorbed by sample of interest under the influence of external DC or Quasi DC magnetic field. Simply speaking, the spectrometer should consist of at least an EM wave excitation source, detector, and transmission line which bridges sample and EM source. The precession frequency of ferromagnetics lies at the regime of microwave (μ-wave) ranged from 0.1 to about 100 GHz, therefore, FMR absorption occurs at μ-wave range. The generation, detection and transmission at such this high frequency are not as simple as those for DC or low frequency electronics. According to transmission line and network theories [10,11], impedance of 50 Ω between transmission line and load has to be matched for optimization of energy transfer. FMR spectrometer also has a resonator and an electro magnet which produces magnetic field to vary the sample's magnetization during the measurement. Sample which is mounted inside the cavity absorbs energy from the μ-wave source. The detector electronics records the changes on either the reflectance or transmittance of theμ -wave while magnetic field is being swept [12]. Most commercial spectrometers, for examples, ELEXSYS-II E580, Bruker Co [13], and JES-FA100, JOEL Ltd. [14], character FMR by measuring the reflectance at fixed band frequency. μ-wave is generated by a Klystron, which goes to a metallic cavity via a wave guide. Signal reflected back from the cavity through the same waveguide to a detector, the Schottky barrier diode. Noted that a circulator is employed, such that the μ-wave does go to the detector and not return back to the generator. The integration of the source, detector, circulator, protected electronics, etc. in a single box, is named "Microwave Bridge". The basic configuration of most FMR spectrometer is shown in Fig.1. FMR absorption spectrum is obtained by comparing the incident and reflected signals. However, the signal to noise ratio (SNR) given by this kind of reflectometer is still not large enough to recognize the spectrum, and lock-in technique is needed to enhance SNR to acceptable value, for example, bigger than 5.

The operational frequency of the source, detector and cavity cannot be varied broadly, therefore, it is necessary to change the microwave bridge and cavity while working with different band frequency. Dart a glance at a FMR spectrometer, the key parts are the microwave bridge, cavity, gauss meter, electromagnet, and lock-in amplifier for signal process. Surely, automation and data acquisition are also crucial.

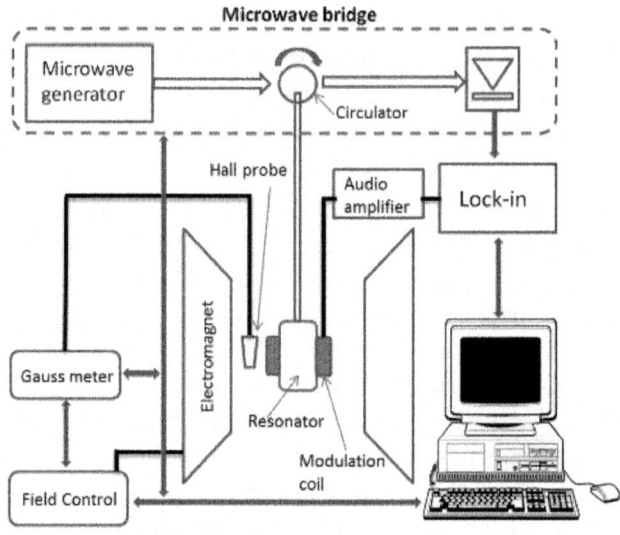

Figure 1: Basic configuration of FMR spectrometer. The microwave bridge mainly consists of the microwave generator, circulator, and detector.

KEY COMPONENTS OF FMR

Metallic Cavity

A Metallic resonator (cavity) is a space enclosed by metallic walls which sustain electromagnetic standing wave or electromagnetic oscillation. The cavity has to be coupled with external circuit which provides excitation energy. On the other hand, the excited cavity supplies energy to the sample of interest (loading) through coupling [10,11]. The basic structure of a cavity set is sketched as inFig.2(a), which consists of a wave guide, a coupler with iris control, and a metallic cavity. The working principle of coupled cavity can be understood by using LRC equivalent circuit which couples to a transmission line [10,11] as shown in Fig.2(b) and 2(c). The coupling structure is represented by an ideal transformer with transforming ratio 1:n. The parallel RLC circuit could be transformed to T_s from AB plane via the transformer, such that C'=n²C, R'=R/n², and L'=L/n². The resonance frequency, f_{res}, and Q-factor will not be affected by the transformation [10] as pointed out by eq.(1)and eq.(2), respectively:

$$\omega_0' = \frac{1}{\sqrt{L'C'}} = \frac{1}{\sqrt{\frac{L}{n^2}n^2C}} = \omega_0$$

(1)

$$Q_0' = \omega_0'C'R' = \omega_0 n^2 C \frac{R}{n^2} = \omega_0 CR = Q_0$$

(2)

A coupling coefficient, β, which defines the relation between energy dissipation (E_d) in a cavity and energy dissipation of external circuit (E_e), tells if the cavity is coupled with external circuit.

$$\beta \equiv \frac{E_d}{E_e} = \frac{\frac{U^2}{2Z_C}}{\frac{U^2}{2R/n^2}} = \frac{R}{Z_C n^2} = \frac{R'}{Z_C}$$

(3)

U in eq.(3) stands for the voltage applied to the cavity, and 1:n is the transforming ratio for tuning to critical couple state ($Z_C = \Rightarrow \beta = 1$) before doing FMR measurement. The characterization impedance, Z_C, is always designed to be 50Ω (Z_0) for matching impedance as mentioned previously. Signal will be distorted and weakened if working at either under-coupling ($\beta<1$) or over-coupling states ($\beta>1$). The external Q factor (Q_e) of a cavity, coupled to external circuit, states the energy dissipation in the external circuit, and also has the following relationship with β:

$$\beta = \frac{Q_0}{Q_e}$$

(4)

In eq.(4), Q_0 is the intrinsic Q factor of cavity. Practically, the loaded Q-factor, Q_L, is used in experiments:

$$Q_L = Q_0 \frac{\beta}{1+\beta} \text{; or } \frac{1}{Q_L} = \frac{1}{Q_0} + \frac{1}{Q_e}$$

(5)

Metallic cavities used for FMR measurement have to be TE mode, such that the cavity center which is also the sample location has maximum excitation magnetic field. Furthermore, this kind of cavity has very high Q-factor and plays key role in FMR detection. The signal output, VS, from the cavity can be expressed as [15]:

$$V_S = \chi'' \eta Q_L \sqrt{PZ_0}$$

(6)

χ'', η, P and Z_0 in eq.(6) stand for imaginary part of magnetic susceptibility of the sample, filling factor, microwave power, and characteristic impedance, respectively. The Q-factor of a cavity directly relates to the detecting sensitivity, which depends very much on the design and manufacturing technique. A good metallic cavity normally has unloaded Q-factor of more than 5,000 which cannot be changed after being manufactured, and therefore is regarded as a constant in eq.(6). Z_0 of 50Ω is fixed for matching the network impedance. We cannot play too much on χ'' and η as well, since the former is the intrinsic behaviour of the sample of interest to be tested, and the last parameter is the volume ratio of the sample and cavity. For thin film and multilayer samples, η is very small, hence this term cannot contribute too much to VS. The source power, P, at the first glance, is the only possible adjustable parameter for sensitivity. Due to the occurrence of saturation, higher power may not be that helpful in increasing the FMR signal. Large excitation power drives magnetization precession in non-linear region which complicates the spectrum analysis. Besides, signal will be reduced and broadened while operating at saturation region. In order to determine the line shapes and widths precisely, this region should be avoided, hence low power is a good choice. The determination of saturation power is not difficult, since the signal intensity grows as the square root of power as indicated by eq.(6). Checking the signal intensity to see if eq. (6) is still validated as the power is increased. It is noted that the maximum excitation power of a common VNA is just a few mWs, and VS is not that clear to follow $\sqrt{P}$.

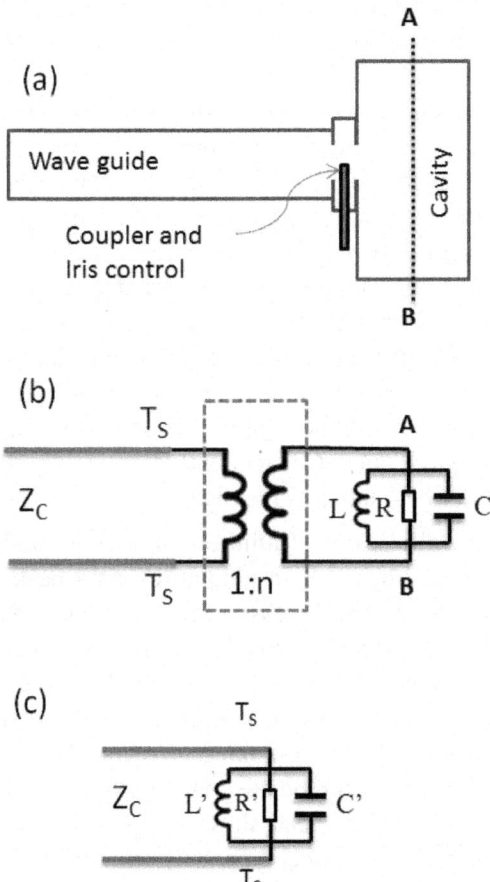

Figure 2: a) The layout of a resonant cavity, (b) The equivalent LRC circuit. The iris control is regarded as a transformer, and (c) The equivalent circuit transformed to the T_s plane [10].

The Q-factor of commercial or home-made cavities can be determined easily by using a VNA, which can also be expressed as:

$$Q = \frac{f_{res}}{\Delta f}$$

(7)

f_{res} is the resonant frequency of a cavity, and Δf, is the full width at half maximum (FWHM) of the resonant peak. Eq.(7) can be simply found by a built-in function of VNA which measures directly the reflected power of the cavity in dB at f_{res}. Besides with dB unit, FWHM (Δf) is also obtained at half power point, i.e. the power level at -3dB. As the values of f_{res} and FWHM are known, Q is determined as shown in Fig.3(d).

Q will be decreased once a sample is inserted. This is because the inserted sample and its holder absorb energy and change the coupling conditions resulting in Q reduction and resonance frequency shift. However, these deviations can be amended a little bit by adjusting the iris which controls the effective impedance by varying the aperture size between the cavity and wave guide, and by adjusting the position of metallic cap. This is equivalent to change the transformer ratio, that is, the equivalent inductance and capacitance. The iris control is a plastic screw (low μ-wave absorption material) which has a gold coated metal cap on one end, and the effective impedance of the whole (wave guide + iris + cavity) depends on the size of the aperture and location of the metal cap. The function of the iris acts as a device to tune the L and C, such that the cavity is critical couple again. As field is swept, the variations of sample's magnetization break the coupling condition so the cavity is deviated from critically coupled state. Consequently, wave is reflected back to the m-wave bridge, and FMR signal is resulted.

For an ideal metallic cavity (infinite conductivity and perfectly smooth inside wall surface), there will be a unique resonant peak with infinite large Q-factor, However, this is not the case in practice. Since the cavity itself has finite conductivity, and further, the inside wall is not perfectly flat, these cause the existence of many peaks with very low Q values as extracted by a VNA shown in Fig.3(a). Fig.3(b)shows the simulation of a copper made cavity with a very rough inside wall. As the roughness is reduced, the spectrum is clearer as shown in Fig. 3(c). This could be due to the μ-wave which is multi-scattered by lumps on the wall surface, and these lumps could also enhance power dissipation further. In consequence, there exists many low Q peaks. Even there are many peaks other than the eigen frequency, these peaks do not response to the change of magnetization, and hence useless for FMR characterization. The manufacture of metallic cavity with Q-factor of higher than 3,000 is laboring. This is because the inside wall has to be mirror polished and coated with a few μm thick Au layer to reduce the imperfection.

Fig.3(a) and Fig.3(d) are the frequency response of a Bruker X-band cavity (TE$_{102}$, ER 4104OR) exhibiting in larger and smaller frequency windows, respectively. The unloaded Q-factor of this cavity is ~9,000 as claimed by the manufacturer, however, our measurement tells that the Q factor is more than 14,000. This is because frequency resolution of the VNA is very high. As the resolution is decreased, Q is found to be decreased as well.

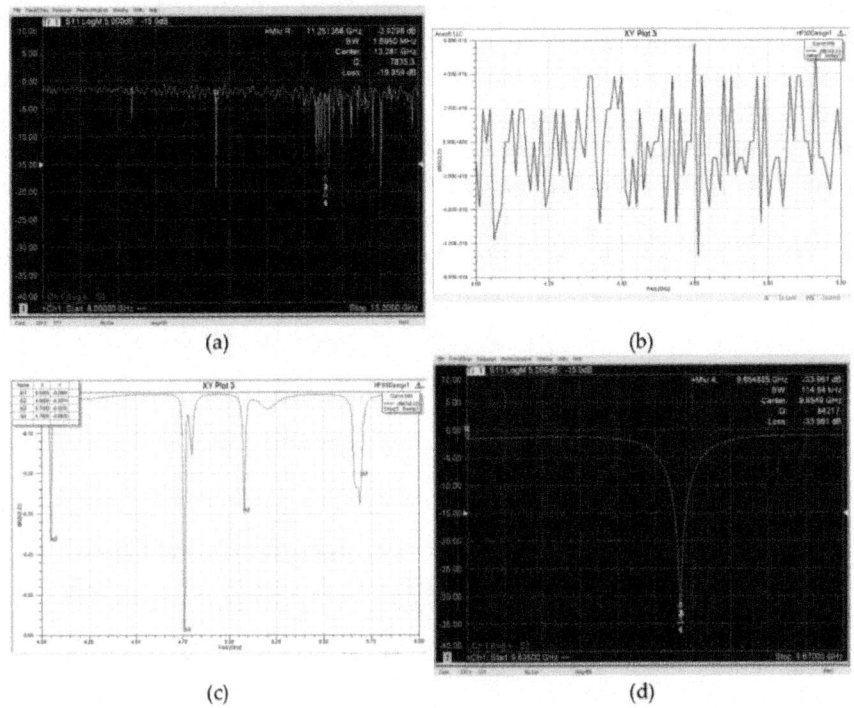

(a)   (b)

(c)   (d)

Figure 3: a) the frequency response of a X-band cavity. Simulation result of a Cu cavity with (b) a very rough inside wall surface, and (c) a smooth inside wall surface. (d) VNA measurement of an X-band cavity with Q higher than 87, 000. Note that there are many side peaks but not appears in this frequency range.

Shorted Waveguide Cavity

The fabrication of metallic cavity with very high Q-factor is not easy as mentioned above. If the lossy of the sample is not very big, cavity with Q of a few hundred to thousand may be good enough to recognize FMR. If so, a shorted waveguide cavity (SWC) could be an alternative. This kind of cavity is just a section of waveguide sealed by a metal plate at one end [16] as show in Fig.4. Since metallic wave guide are commercially standard with wide range of frequency from L to W band, it is worth to obtain FMR information at different bands with this cheap, simple and effective method. If the length of the waveguide tube is equal to $n\lambda/2$, standing wave can be formed inside. In that n and λ represents for integer and wave length of the μ-wave, respectively. In order to excite FMR signal, sample has to be placed at the $n\lambda/4$ position away from the shorted plate at which maximum magnetic field is located. In Fig.4, n=0 and n=1 are the suitable locations, however, n =0 gives advantages of easier sample manipulation, and lesser interference from irrelevant insertions. Sample is mounted at the top of a plastic screw through the shorted plate, such that its position can be adjusted for maximum signal. However, SWC does not have an iris to tune the impedance, hence critical coupling may not be easy to obtain. Although the Q is somewhat lower (less than 2,000), SWC is easy to make from commercial waveguide tube. Furthermore, different band frequency cavity can be built in the same manner, and this is particular convenient to study FMR at different band frequency with network analyzer.

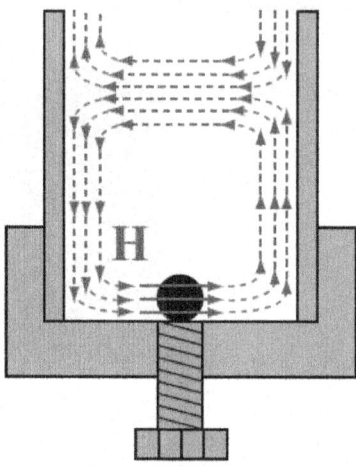

Figure 4: A shorted wave guide with possibility of sample manipulation. Magnetic field distribution originated from the μ-wave is also sketched [16].

Microstrip and Co-Planar Waveguide

Microstrip (MS) and co-planar waveguide (CPW) which are indeed transmission lines, are commonly used to extract FMR at broad frequency range [16,17,18,19]. Sample is mounted on top of their signal lines, and μ-wave is conducted into MS (CPW) via Port 1 of the VNA, and the differential change in either reflected or transmitted signal is analyzed. In this operation scheme, frequency is swept at fixed field, and once the FMR conditions are fulfilled μ-wave absorption occurs. Fig.5 and Fig.6 are the sketches of MS and CPW. Since $h \ll \lambda$, μ-wave propagates in these two lines is regarded as quasi-TEM mode. In order to match the impedance of 50Ω, the dimensions of these two planar transmission lines are restricted.

In case of MS, we have the followings [11]:

The effective dielectric constant, ε_{eff} has approximately the form:

$$\varepsilon_{eff} = \frac{\varepsilon_r+1}{2} + \frac{\varepsilon_r-1}{2}\frac{1}{\sqrt{1+12h/w}}$$

$$(8)$$

The dimension of the strip line and characteristic impedance, Z_c, can be worked out as below:

$$Z_c = \begin{cases} \frac{60}{\sqrt{\varepsilon_{eff}}}\ln\left(\frac{8h}{w}+\frac{w}{4h}\right) & ; \ \text{for } w/d \leq 1 \\ \\ \frac{120\pi}{\sqrt{\varepsilon_{eff}}\left[\frac{w}{d}+1.393+0.667\ \ln\left(\frac{w}{d}+1.444\right)\right]} & ; \ \text{for } w/d \geq 1 \end{cases}$$

$$(9)$$

In case of CPW if the thickness of signal line is ignorable, then [10]:

$$\varepsilon_{eff} = \frac{\varepsilon_r+1}{2}$$

$$(10)$$

$$Z_c = \begin{cases} \frac{\eta_0}{\pi\sqrt{\varepsilon_{eff}}}\ln\left(2\sqrt{\frac{b}{w}}\right) \ \Omega & ; \ 0 < w/b < 0.173 \\ \\ \frac{\pi\,\eta_0}{4\sqrt{\varepsilon_{eff}}}\left[\ln\left(2\frac{1+\sqrt{w/b}}{1-\sqrt{w/b}}\right)\right]^{-1} \ \Omega & ; \ 0.173 < \frac{w}{b} < 1 \end{cases}$$

$$(11)$$

η_0 in eq.(11) stands for the wave impedance in free space. The $Z_c = Z_0$ is set to 50Ω, then b, d and wcan easily be determined from eq.(9) and eq.(11). The line width of the strip can be ranged from sub-millimeter to millimeters depending on the design and dielectric material used as implied by these equations. The size of this kind of line width can be fabricated simply by photo lithography together with life-off process. Due to the large loss, the characterization of nano size sample may not be easy and lock-in technique is needed for SNR enhancement as described in next section.

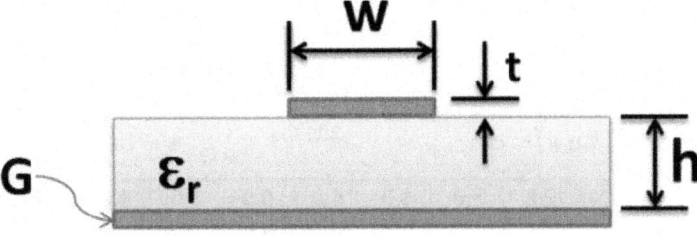

Figure 5: The geometry of a microstrip line, where εr is the relative dielectric constant of the substrate.

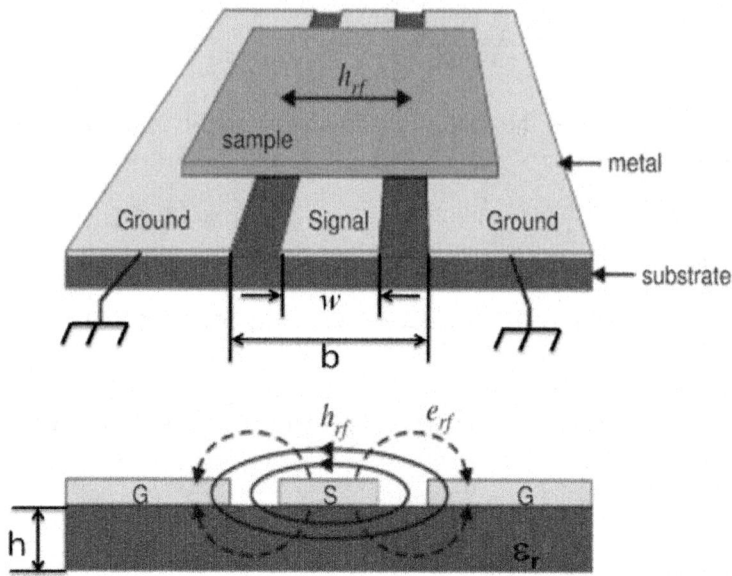

Figure 6: Schematical diagram of a coplanar waveguide along with the dimension. The lower diagram shows the magnetic (full lines) and electric (dashed lines) field lines winding around the signal conductor (S).

FMR can determined by characterizing the transmittance, ie. S_{12}. However, if shorted MS, and CPW are used, S_{11}, the reflectance FMR can also be found. Fig.7 shows the FMR of permalloy films extracted by VNA and CPW [18].

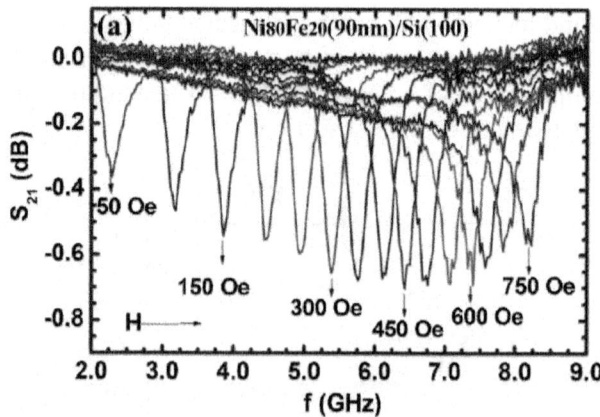

Figure 7: A typical FMR transmittance spectrum of permalloy films extracted by VNA with CPW [18].

TECHNIQUE FOR SIGNAL TO NOISE ENHANCEMENT

Lock-in Phase Detection

If signal is deeply buried inside noise floor, lock-in method is normally employed to enhance the signal to noise ratio (SNR) by narrowing the detecting band window [20]. Considering a DC Signal, S, which is heavily contaminated by white noise, n, and for every measurement comes out from a measurement equipment, a voltmeter for an example, the reading can be expressed as:

$$V = S + n$$

$$(12)$$

If S can be modulated sinusoidally at a certain frequency somehow, then:

$$V(t) = S\cos(\omega t) + n$$

$$(13)$$

Note in eq.(13) that the modulation of white noise is still a white noise.

There is also a reference signal with same frequency but could be different in modulating amplitude (V_a) and phase (φ), i.e.,

$$V_{ref} = V_a \cos(\omega t + \phi)$$

$$(14)$$

The product of eq.(13) and (14) gives:

$$V(t)V_{ref} = SV_a \cos(\omega t)\cos(\omega t + \phi) + nV_a \cos(\omega t + \phi) =$$
$$\frac{SV_a}{2}[\cos(\phi) + \cos(2\omega t + \phi)] + n\,V_a \cos(\omega t + \phi)$$

$$(15)$$

The second and third terms can be eliminated if these two parts are passed to a low pass filter with cutoff frequency setting at $\omega/2$ or lower. Finally, we have:

$$V \propto \frac{SV_a}{2}\cos(\phi)$$

$$(16)$$

That is, the signal is amplified by V_a, and has a maximum while φ is "phase-locked" to zero, hence the name of lock-in amplifier.

Feld Modulation Lock-In Technique and Derivative Spectrum

It is easier to determine H_{res}, and ΔH_{pp} by differentiating the original signal. To do so, field modulation lock-in detection is employed and described below.

There is a small field, H_a, with modulation frequency of ω superposing on top of external DC magnetic field, H_0:

$$H(t) = H_0 + H_a \cos(\omega t) \tag{17}$$

Assuming we have FMR signal, V_{FMR}, whose Taylor expansion at H_0 is given below:

$$V_{FMR}(H) = V_{FMR}(H_0) + \left.\frac{dV_{FMR}}{dH}\right|_{H=H_0} H_a \cos(\omega t) + \cdots \tag{18}$$

Meanwhile, we also have another signal, the reference, which has the same frequency as that of the modulation field, but with a phase angle, φ:

$$V_{ref} = \cos(\omega t + \phi) \tag{19}$$

The product of eq.(18) and (19) gives eq.(20) below:

$$V_{FMR}(H) \times V_{ref} = V_{FMR}(H_0)\cos(\omega t + \phi)\left.\frac{dV_{FMR}}{dH}\right|_{H=H_0} + H_a\cos(\omega t)\cos(\omega t + \phi) + \cdots =$$
$$V_{FMR}(H_0)\cos(\omega t + \phi) + \frac{1}{2}\frac{dV_{FMR}}{dH}H_a\cos(\phi) + \frac{1}{2}\frac{dV_{FMR}}{dH}H_a\cos(2\omega t + \phi) + \cdots \tag{20}$$

The first and third terms in eq.(20) can again be removed by using a low pass filter with cutoff frequency setting at $\omega/2$ or even lower. The second term is time independent, proportional to derivative of the input signal and magnify by H_a, which has a maximum if the phase, φ, is locked to 0. Another advantage of use this method is that 1/f noise and drift problem can be excluded by setting the sampling window at high frequency.

The way to turn the DC or quasi DC signal from to AC is an important subject. The simple and normal way to do this is to insert two coils which sandwich the cavity. These coils are driven by a power amplify at certain high frequency, such that an AC magnetic field of a few mT at a frequency up to 200 kHz can be produced. Therefore, the output signal consists of the modulation and DC components. All these together with the reference are sent to the lock-in amplify for SNR enhancement and derivative spectrum as mentioned previously. Modulation amplitude (MA), modulation frequency (MF), and time constant (TC) which is the reciprocal of the low pass filter's cutoff frequency, have large influence on the spectra. Signal strength is linearly proportional to the MA while MA is small. Simply speaking, MA should be small enough for sampling in linear region, but needs to be large enough for gaining good sensitivity. However, too large the MA results in signal distortion as shown in Fig.8(a). This can be understood from eq.(20) that high order term cannot be ignored for large MA which causes distortion of derivative signal. The choice of MF is critical as well that small MF cannot get rid of 1/f noise completely, and elongates the acquisition time. Large MA and MF also cause passage effect that the rate of "passage" through the absorption line is faster than the relaxation rates, which results in distorted spectrum, or even inversion of signals upon reversal of the field scan direction. Therefore, it is important to check the FMR signal with either a standard or a well-known sample with different lock-in conditions for the best settings. In order to obtain correct line shape spectrum, the rule of thumb is to set the MA about 1/10 of the FWHM and increases to about 1/3 if necessary. TC also needs to coordinate with MF. According to Nyquist sampling theorem, the sampling rate should be at least twice the highest frequency contained in the signal [21]. For example, if the MF is 100 kHz, the sampling frequency would be at least 200 kHz at which TC is about 5µS. If this restriction does not fulfill, sampling points cannot not be captured immediately. Consequently, information loss and line shape distortion are always resulted. The best way to avoid this is to set the scan time for a FMR signal 10 times longer than the time constant.

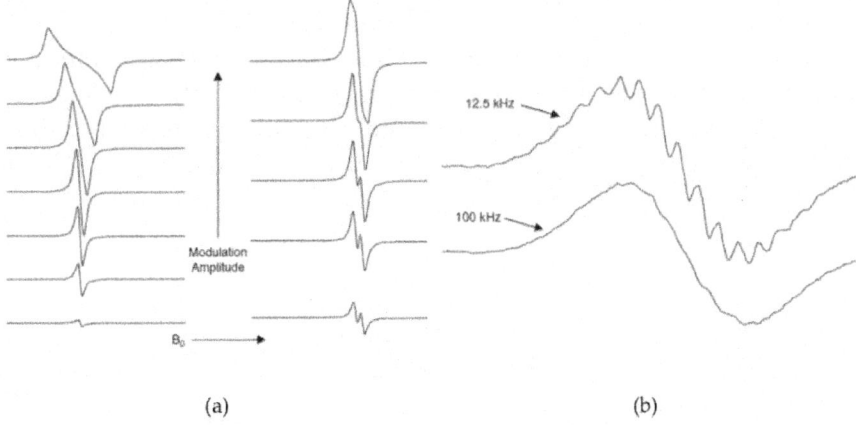

(a)   (b)

Figure 8: The influence of modulation amplitude (a) and modulation frequency (b) on signal. Distortion will be resulted if MA and MF are not appropriate [15].

VECTOR ANALYZER BASED FMR SPECTROMETER

Basics of Vector Network Analyzer (VNA)

VNA is indeed an instrument developed for characterization of electrical devices (device under test, DUT) by sending an electromagnetic wave. As an analogue of that in optics, the incident wave (either optical or micro wave) will be reflected or/and transmit after interacting with the DUT. By examining the reflectance and transmittance, that is, the ratios of the powers of reflected, transmitted to that of the incident waves, scattering parameters of the DUT can be found as depicted in Fig.9. VNA cannot only find out the reflectance and transmittance, but also the impedance, phase lag, insertion loss, return loss, voltage standing ratio, etc., can be worked out. However for FMR experiment, only the first two functions are employed. VNA has at least two ports, and each port can produce and measure μ-wave. The scattering parameter S_{ij} stands for scattering power ratio of incident wave produced by port i, and measured by port j. That is, for a two-port VNA, S_{11} and S_{22} characterize wave reflectance, while S_{12} and S_{21} determine the transmittance. If FMR is determined by reflected spectrum, one port is enough by analyzing either the S11 or S22. For transmittance spectrum, two ports are required for either the S_{12}, or S_{21} determination. VNA has capability of microwave generation and detection, and furthermore, nowadays model has frequency ranged from about 0.1 to 100 GHz, or even higher at excitation power of tens dBm, Thus, it can serve as a microwave bridge to extract FMR parameters at a very broad frequency band. VNA measures not only the scattering amplitudes, but also their correlated phases. This is in contrast to its counterpart, the scalar network analyzer (SNA) which cannot tell phase information. FMR signal is extracted from power absorption which contains no phase information, and SNA should be good enough for the purpose. However, this kind of machine has been obsoleted for many years.

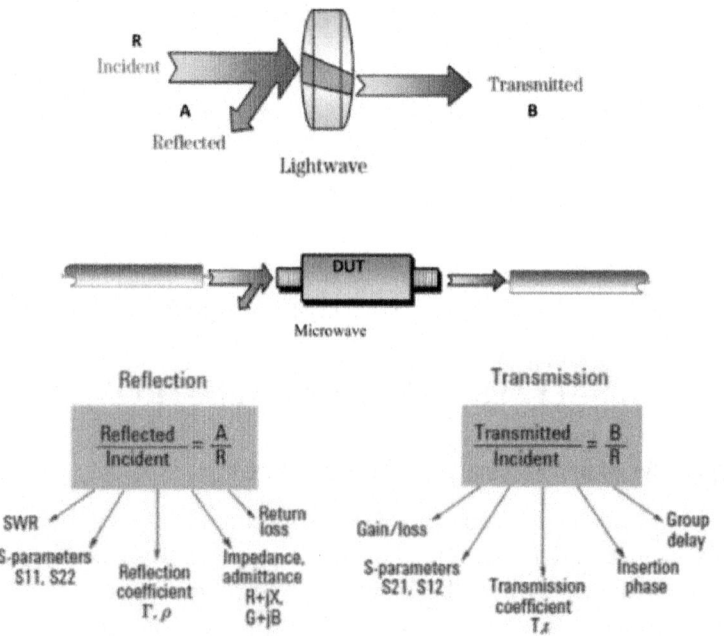

Figure 9: The fundamentals of microwave network analysis are analogue to that of the optics. The lower part of the panel indicates the definitions of scattering parameters [24].

VNA is commonly used with MS and CPW for FMR determination with frequency swept at fixed field. Since signal output from VNA cannot be plugged into lock-in amplifier directly, the characterization of nano scale sample is thus difficult without using lock-in. The application lock-in with VNA will be discussed in next section.

Field Swept VNA-FMR

Metallic cavity usually has rather higher Q-factor and could be simply used with a VNA for field swept FMR detection. Taking the advantages of high Q cavity and VNA, C.K. Lo, *et. al.* [23] demonstrated FMR measurement without employing field modulation lock-in technique. They employed a TE_{102} cavity with unload Q of ~9,000 at X-band, and FMR signal was recorded through the measurement of reflectance, and therefore one VNA port for S_{11} was used. This powerful method extracts signal easily and directly as shown in Fig.10(a). One could if necessary, differentiate the original data for derivative spectrum which is used to determine the peak position and line wide as those in commercial FMR spectrometer. Further, this combination has quite good sensitivity that 1.6 nm CoFeB can be detected with SNR better than 5 as shown in Fig.10(b).

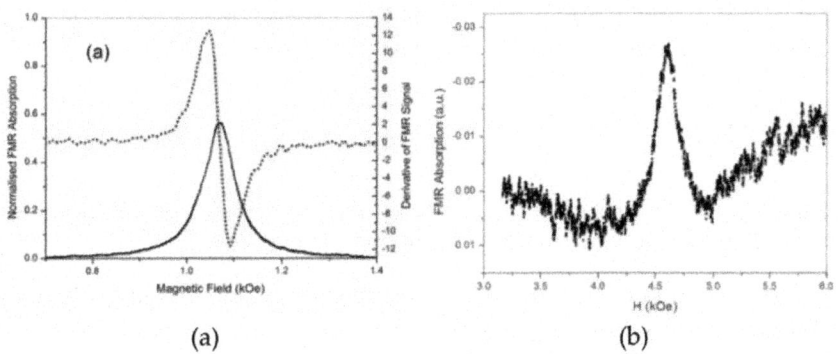

(a)   (b)

Figure 10: a) FMR of 5 nm Py exacted at P = 10 dBm. The dotted line is the derivative of experimental data. (b) Signal extracted from a 1.6nm CoFeB.

A new built spectrometer should be characterized before properly used and well known samples, such as permalloy, Fe, Co, etc., ferromagnetics are normally employed due to their ΔH and peak positions can be found widely in literatures. Also, these samples are easy to prepare with different thicknesses, and results should be comparable to those in literatures reported.

There are many build-in useful functions with nowadays VNA, and only some of them are used, for instances, the traces of valley (dip) position, Q factor, band width, average, etc. Dip frequency and Q-factor will be varied as the sample's magnetization state is changed by external field. The changes of these quantities were also recorded simultaneously for a Fe/Ag multilayer as shown in Fig.11 in which (a) is the FMR absorption, and (b), the post derivative of (a). The shifts of resonant frequency and Q-factor are shown in (c) and (d), respectively. Despite the variation of center frequency is just a few hundreds MHz, it allows us to determine ferromagnetic parameters precisely. The change of Q-factor is upside-down to that of the absorption. This is because any absorption of microwave inside the cavity will reduce the Q value.

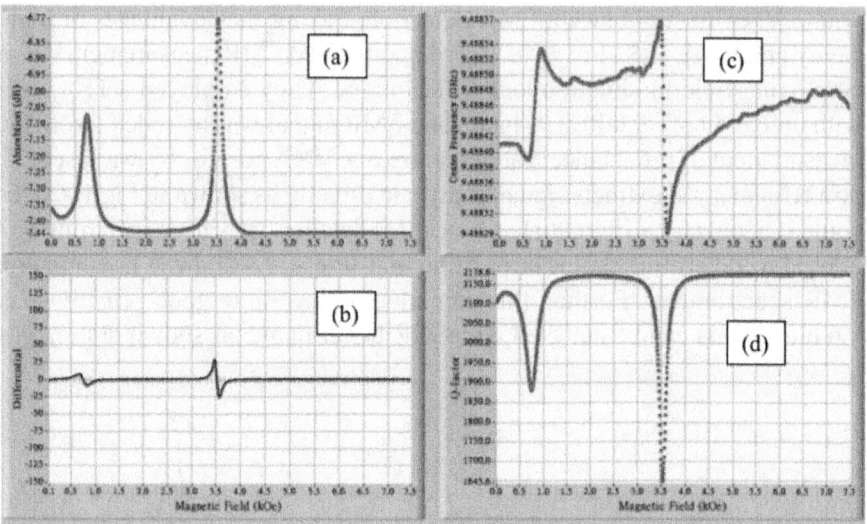

Figure 11: The changes of FMR (a), resonance frequency (c), and Q-factor are recorded. (b) is the derivative spectrum taken from (a) mathematically.

The excitation power of VNA is only tens dBm, and it is unlikely to drive the sample into non-linear regime. Signal intensity is proportional to $\sqrt{P}$ as indicated by eq. (6), however, SNR is found to be not much different if P is bigger than -20 dBm as demonstrated in Fig.12.

As mentioned previously that VNA has very high frequency resolution, apparently, Q could be tuned as high as 140,000 and more by simply adjusting the iris. However, distorted spectrum is resulted at very high Q, and two peaks appear while Q is adjusted close to maximum Q as seen in Fig.13(a) and (b), respectively. It is also found that the SNR does not varied significantly if Q lies between 2,000 to 13,000 (the unloaded Q is about 9,000 as claimed). Surely, this finding is just for referral, and would depend very much on cavity used.

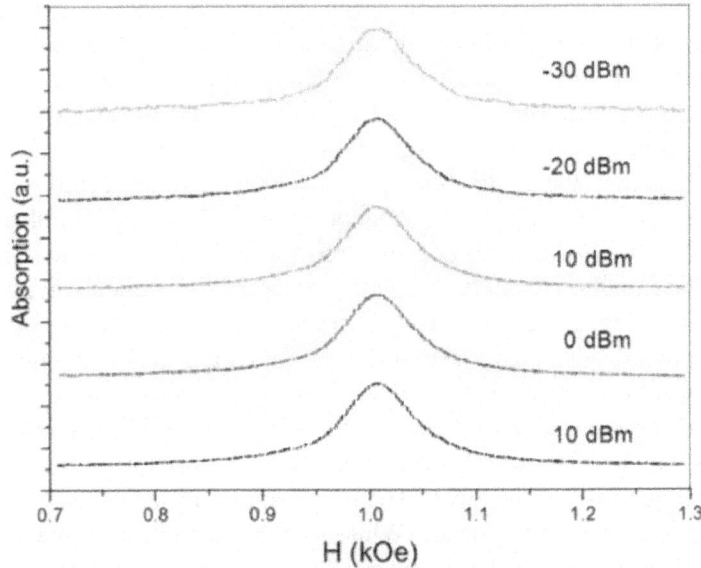

Figure 12: FMR signal of 10nm Py was recorded as function of excitation power. Note that these spectra have the same scale, but different offset for clear comparison.

The advantages of the VNA-FMR with high Q cavity are: (1) the sensitivity which is comparable to that of commercials, but a lot cheaper, (2) multi frequencies can be done in the same manner by just changing to another band cavity only within the frequency range of the VNA.

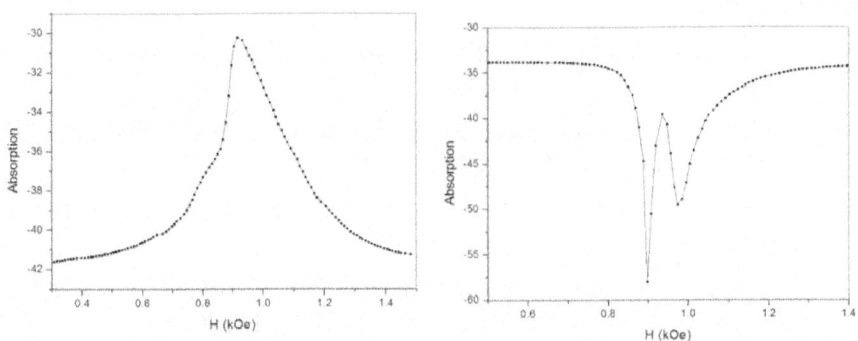

Figure 13: a) data taken at $Q \approx 50,000$, and (b) at ~110,000. Clearly, distorted signals are obtained.

Nevertheless, some points needed to be well aware. Firstly, VNA has built-in sweep average function which increases the SNR for every sweeping. The penalty of using high average count is the very long acquisition time.

For example, a commercial FMR spectrometer takes about a few seconds to sweep a spectrum over 2 kOe field at 1 Oe resolution with time constant of a 1mS. However, the settled time for a VNA with average of 1 is found to be about 0.1S a point. Therefore, it takes at least 3 min to obtain the spectrum with the same conditions. Average of 10, and sometime 50 are often used for better SNR, then the acquisition time is very long which could overheat the electromagnet. Secondary, VNA cannot work with lock-in amplifier directly, and a circulator is also needed to force the reflected wave to a new port. Before the signal is conducted to the lock-in amplifier, a microwave transducer (Schottky diode, for example) is required. This is because the output from VNA is power which has to be transduced into voltage before plugging into the lock-in amplifier. Besides, the output of VNA has already been averaged over a range of frequencies. Thus the Schottky detector cannot recognize this tiny average change. This problem can be amended by narrowing the starting and ending frequencies of the VNA. The two frequencies indeed can be set to the same value as the dip frequency. While working with lock-in, the VNA-FMR acquires data as fast as the commercial one, since the SNR enhancement is processed by the lock-in and not by the VNA. Further, MS-, and CPW- FMR can also work together with lock-in method in this manner. If one wants to study high power VNA-FMR, a high power microwave amplifier could be used. However, if the VNA is still used as the analyzer, a suitable attenuator has to be inserted between the returned line and VNA, otherwise damage is resulted.

REFERENCES

1. M. Di, az de Sihues, C. A. Durante-Rincon, and J. R. Fermin, J. Magn. Magn. Mater. 316, e462 (2007

2. V. G. Gavriljuk, A. Dobrinsky, B. D. Shanina, and S. P. Kolesnik, J. Phys, Condens. Matter 18, 7613 (2006

3. Z. Zhang, L. Zhou, P. E. Wigen, and K. Ounadjela, Phys. Rev. B 50(9), 6094 (1994

4. B. Heinrich, Y. Tserkovnyak, G. Woltersdorf, A. Brataas, R. Urban, and G. E. W. Bauer, Phys. Rev. Lett. 90(18), 187601 (2003

5. V. P. Nascimento, and E. Baggio, F. Saitovitch, L. C. Pelegrinia, A. Figueiredo, E. C. Biondo, Passamani, J. Appl. Phys. 99, 08C108 (2006

6. K. Lenz, E. Kosubek, T. Tolinski, J. Lindner, and K. Baberschke, J. Phys, Condens. Matter 15, 7175 (2003

7. M. Oogane, T. Wakitani, S. Yakata, R. Yilgin, Y. Ando, A. Sakuma, T. Miyazaki, J. J. Appl, A. Phys, 2006

8. R. Urban, G. Woltersdorf, and B. Heinrich, Phys. Rev. Lett. 87(21), 271204-1 (2009

9. T. Kato, K. Nakazawa, R. Komiya, N. Nishizawa, S. Tsunashima, and S. Iwata, IEEE Trans. Magn., 4411NOVEMBER 2008

10. "Microwave Electronics: Measurement and Materials Characterization"L. F. Chen, C. K. Ong, C. P. Neo, John Wileys & Sons Ltd. 2004

11. "Microwave Engineering", D.M. Pozar, John Wileys & Sons Ltd. (2012)

12. "High-Frequency EPR Instrumentation"E.J. Reijerse, Appl Magn Reson (2010

13. See http://wwwbruker-biospin.com/epr-products.html.

14. See http://wwwjeol.com/PRODUCTS/AnalyticalInstruments/Electron SpinResonance/tabid/98/Default.aspx.

15. "Quantitative EPR"G. R. Eaton, S. S. Eaton, D. P. Barr, and R. T. Weber, Springer Wien, New York, (2010

16. Marina VroubelYan Zhuang, Behzad Rejaei, and Joachim N. Burghartz, J. Appl. Phys. 99, 085062006

17. C. Nistor, K. Sun, Z. Wang, M. Wu, C. Mathieu, and Matthew Hadley Appl. Phys. Lett. 95, 012504 (2009

18. Y Chen, , D.S Hung, , Y.D Yao, , S.F Lee, , H.O Ji, , C Yu, , J Appl, . C Phys, 104 (2007)

19. H. G. Lowi, J. Nski, DUBOWIK, Acta Physicae Superficierum, Vol. XII, 2012

20. "The Art of Electronics"P. Horowitz, W. Hill, Cambridge Univ. Press, 2nd Ed. 1989

21. Advanced Digital Signal Processing and Noise Reduction"S.V. Vaseghi, John Wiley & Sons Ltd., 2008

22. Agilent VNA user manual

23. C. K. Lo, W. C. Lai, J. C. Cheng, Rev. Sci. Instruments, 82, 086114 (2011

24. R. W. Damon, Rev. Mod. Phys., 25, 11953

25. N. Bloembergen, S. Wang, Phys. Rev., 93 11953

Chapter 8

A NOVEL INSTRUMENTATION CIRCUIT FOR ELECTROCHEMICAL MEASUREMENTS

Li-Te Yin[1], Hung-Yu Wang[2], Yang-Chiuan Lin[2] and Wen-Chung Huang[3]

[1]Department of Optometry, Chung Hwa University of Medical Technology, Tainan 717, Taiwan

[2]Department of Electronic Engineering, National Kaohsiung University of Applied Science, Kaohsiung 807, Taiwan

[3]General Education Center, Chung Hwa University of Medical Technology, Tainan 717, Taiwan

ABSTRACT

In this paper, a novel signal processing circuit which can be used for the measurement of H^+ ion and urea concentration is presented. A potentiometric method is used to detect the concentrations of H^+ ions and urea by using H^+ ion-selective electrodes and urea electrodes, respectively. The experimental data shows that this measuring structure has a linear pH response for the concentration range within pH 2 and 12, and the dynamic range for urea concentration measurement is in the range of 0.25 to 64 mg/dL. The designed instrumentation circuit possesses a calibration function and it can be applied to different sensing electrodes for electrochemical analysis. It possesses the advantageous properties of being multi-purpose, easy calibration and low cost.

INTRODUCTION

The prototype of biosensors was first proposed by Clark and Lyon in 1962 [1]. The analytical method of detecting organisms actually exploits the molecular recognition between enzyme and acceptor. This concept involves placement of an enzyme in close proximity to an electrode surface, where the enzyme is able to catalyze a reaction. The analysis is based on the measurements of the consumption of an elective reactant (O_2) and the production of an electroactive product (H_2O_2) [2]. The sensing mechanism of the biosensor is dependent on the biological specificity of the enzyme-catalyzed reaction and the selectivity of the ion-selective electrode, and hence, the characteristics of the biosensor are strongly related to the selectivity of the ion-selective electrodes. The enzyme electrode is a miniature chemical transducer which functions by combining an electrochemical procedure with immobilized enzyme activity. In 1967, Updike and Hicks used glucose oxidase immobilized on a gel to measure the concentration of glucose in biological solutions and in tissues *in vitro* [3]. From this moment on, many researches devoted to the development of biosensors, such as the O_2, H_2O_2, H_2, H^+, NH_3, CO_2 electrodes and ion-sensitive field effect transistor (ISFET) [4]. An ISFET can be considered as a special type of the MOSFET without a metal or polysilicon gate, with the gate coated with a hydrogen ion-sensitive layer [5]. The gate of ISFET is directly exposed to the buffered solution to detect the concentration of hydrogen ion. The extended gate field effect transistor (EGFET) is another sensing structure which isolates the FET from the chemical environment.

Biosensors mainly composed of two parts. The first part is the sensing element which receives the input signal for the biological sensor. It can be the organism molecules, tissue or molecular recognition elements of individual cells. The other part is the electronic circuit which processes the quantified electronic signals from sensing element and outputs the processing result [6]. Therefore, the way to get accurate biological information quickly from sensing element and its processing circuit receive much attention of researchers [7,8]. Electrochemical sensors are widely utilized in many applications, such as disease diagnosis, food inspection and environmental monitoring, because of their fast reaction, high selectivity, high sensitivity, and simplicity [9].

In this study, based on the potentiometric method, an electronic instrumentation circuit is designed to detect the concentrations of H^+ ions and urea by using H^+ ion-selective electrodes and urea electrodes. The system performance for the H^+ ions concentration detection can achieve the same accuracy as the commercial pH meter. The urea concentration detection using urea biosensors based on the measurement of H^+ ion concentration possesses the dynamic range between 0.25 and 64 mg/dL. The workability of the sensing system is verified by measurement results.

REALIZATION OF SENSING CONFIGURATION

In this study, the used direct potential method is based on the measurement of potentials of electrode and the analysis of activity concentrations of ions employing the Nernst equation. The method usually uses indicating electrodes with ion-selective function. There are slight structural differences between the electrodes used and their general structure is shown inFigure 1. The SnO_2/ ITO/PET pH electrode in Figure 1 is based on a separated structure [10].

The SnO$_2$ thin film is deposited at a thickness of 200 nm using sputtering [11]. Figure 2 shows the practical electrode connected with a coaxial wire to increase the immunity to external noise. For the measurement of the concentrations of H$^+$ ions, the pH sensing area acting as working electrode (WE) and Ag/AgCl reference electrode (RE) were dipped into buffer solution and connected to the input terminals A and B of the designed instrumentation, as shown in Figure 3. Figure 3 also shows the proposed potential system structure used for the concentration measurement of hydrogen ion. The circuit mainly consists of an 8-bit microprocessor chip module (P89C51RB2HBA, Philips), an analog to digital converter (ADC), a liquid crystal display module (LCM) and a precision voltage amplifier. The voltage amplifier is implemented by an instrumentation amplifier (IA) to make good use of its characteristics of low-noise, high input impedance, low output impedance and tunable gain of the instrumentation amplifier. The commercially available pH meter usually set the zero potential which corresponds to pH 7. One unit change of pH value corresponds to the voltage change of about 59.1 mV. In theory, pH 14 to pH 0 will have the voltage from −413.7 mV to +413.7 mV.

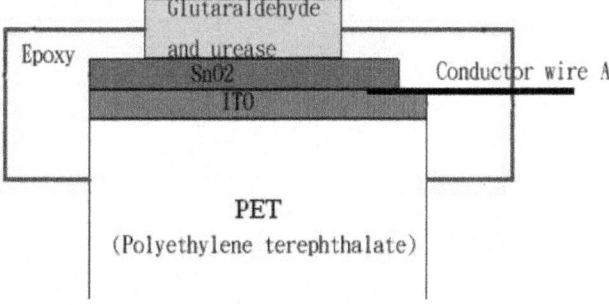

Figure 1: Electrode structure of ion-contact type.

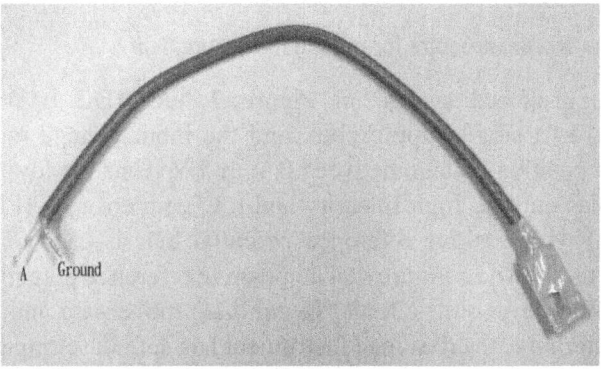

Figure 2: Practical pH electrode connected with a coaxial wire.

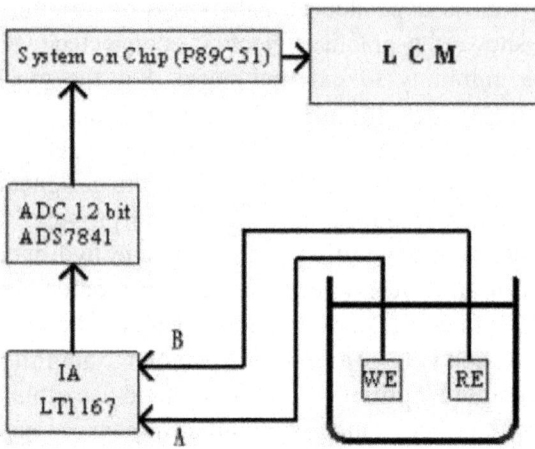

Instrumentation amplifier circuit

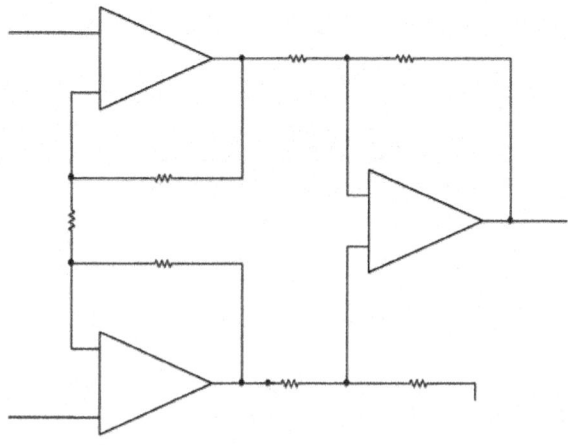

Figure 3: The system structure for potential measurement.

For the proposed system in Figure 3, the ADC (ADS7841, Texas Instruments) is a single-supply chip, and the input voltage range of analog channel is a positive voltage between 0 V to 5 V. Thus the low-offset voltage and input bias current, high linearity and low gain error IA (LT1167, Linear Technology) with positive reference potential bias is used to construct the instrument, as shown in Figure 4. The positive reference potential bias can be obtained with a level-shift circuit. To calibrate the system and maximize the measurement range, the designed instrument has default setting that the output voltage of the IA for potential 2.5 V which corresponds to pH 7 of the buffer solution and code 2048 of ADC.

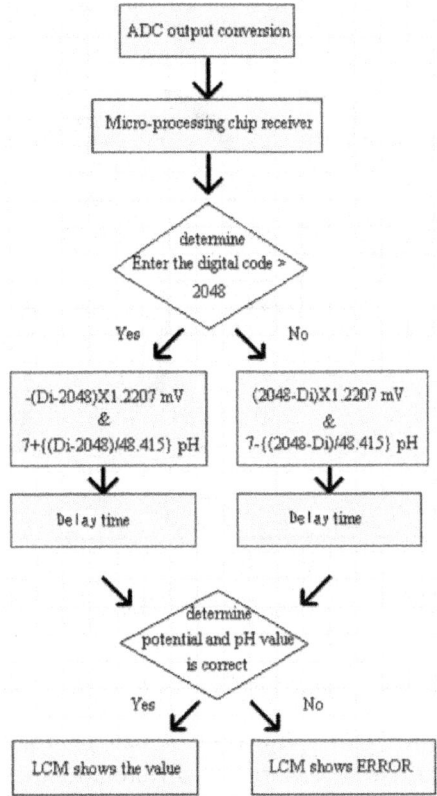

Figure 4: Instrumentation amplifier circuit.

The 12-bit ADC (ADS7841) is operated with a supply voltage of 5 V. Due to its input voltage range of 0 to 5 V, the minimum voltage step can be $5/2^{12}$ V = 1.2207 mV. The default setting is that the change of per unit of pH value corresponds to 59.1 mV potential change, but this value is settable for our system. Therefore, the accuracy of measured pH value can achieve the resolution of about one digit after decimal point. It is enough for the measurement of general chemical laboratory. Figure 5 shows the internal program operation flow chart for P89C51 chip. For the operation procedure in Figure 5, the measured value is obtained using the output code of the ADC multiply by minimum voltage step (1.2207 mV). The pH value corresponding to its measured voltage value can be derived according to the procedure in Figure 5. An adjustable delay time is added to stabilize the displayed output values on LCM since the output codes of ADC (with conversion rate of 200 ksample/s) is refreshed too frequently. Then the measured value is checked and displayed on LCM.

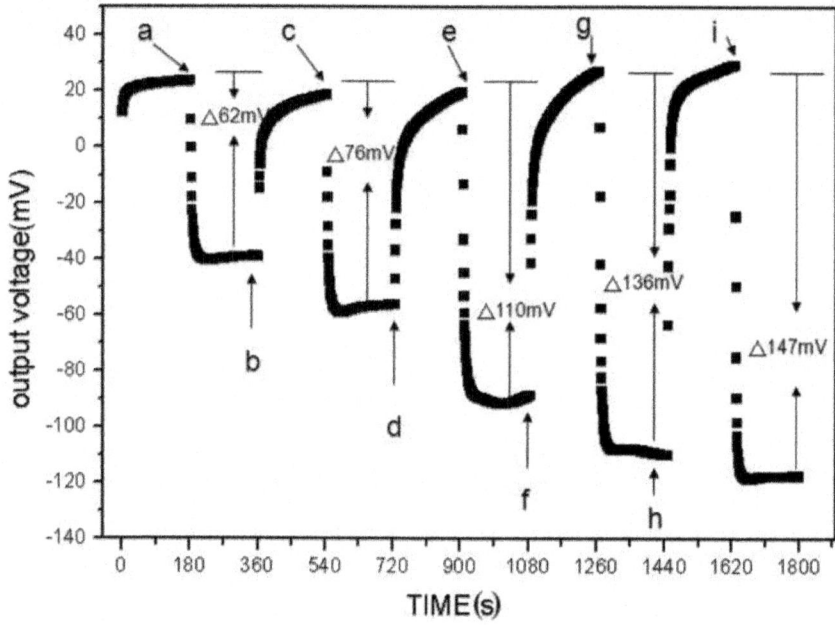

Figure 5: The internal operation of P89C51 for pH measurement.

According to the Nernst equation, it can be found that the measured voltage of a solution will be different at different temperatures. Therefore, the designed instrument has adjustable setting functions. That is, it has default setting that the 2.5 V output voltage of the IA corresponds to the measured voltage of calibrated solution of pH 7 and code 2048 of ADC. However, this corresponding code of ADC is settable. By inputting the program with the correct code of ADC (which is obtained by measuring the buffer solution of pH 7) from the COM port in designed electrochemical sensing instrumentation, the calibration can be attained. Therefore, the temperature calibration function can be achieved by this designed circuit. In addition, the designed program in P89C51 has the reset function. It adopts the mean of 50 output codes of ADC to process when the reset function is triggered.

Moreover, the urea biosensor circuit is constructed. The similar potentiometric method is used for the concentration measurement of urea solutions by using urea electrodes. The urea solution with higher concentration results in higher measured potentials according to:

$$CO(NH_2)_2 + 3H_2O \rightarrow CO_2 + 2NH_4^+ + 2OH^-$$ (1)

The differences between the concentration measurement instrument for H^+ ion and urea are the different sensing electrodes and the internal program of P89C51. For setting the instrumentation, the five urea solutions with concentrations of 8 mg/dL (a), 12 mg/dL (c), 16 mg/dL (e), 24 mg/dL (g) and 32 mg/dL (i) are used for calibration, as shown in Figure 6. It also shows the measured potential difference for different urea solutions. We can observe that the output voltages are increased with the higher concentration of urea solutions.

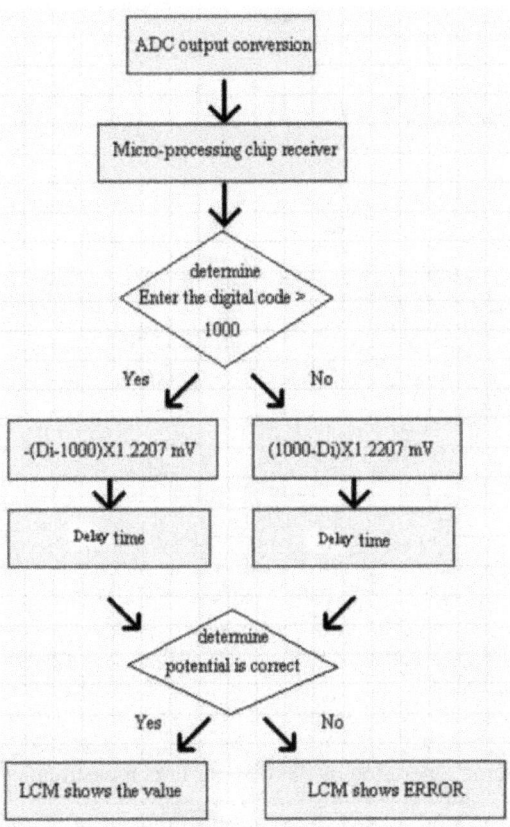

Figure 6: Measurement result of urea concentration.

As shown in Figure 7, the output voltage of 1.2207 V corresponds to output code 1000 of ADC of the instrument because the mentioned minimum voltage step is 1.2207 mV. The electrode is immersed in the phosphate buffer (PB) solution for resetting after each concentration measurement of different urea solutions. However, we observe that the restored voltage is not an exact 1.2207 V after the electrode is immersed in the PB solutions. To solve this problem, the microprocessor is programmed such that the default 1.2207 V which corresponds to output code 1000 of ADC is adjustable.

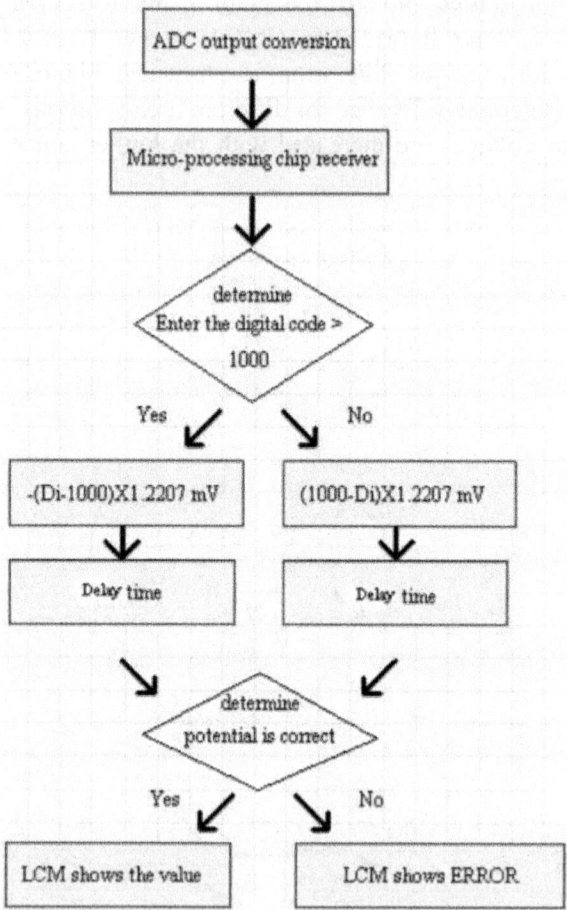

Figure 7: The internal operation of P89C51 for urea measurement.

By inputting the microprocessor with the derived code after the electrode was immersed in the PB solution to replace the default code 1000 of ADC, the reset function is attained. It must be noted that the selected IA, ADC and microprocessor chips for the realization of this designed instrument are based on the considerations of appropriate accuracy, low cost and easy modification and debugging for experimental demands.

EXPERIMENTAL RESULTS

The designed circuit for practical measurement is shown in Figure 8. It is used to measure the concentrations of hydrogen ion and urea by using the H^+ ion-selective electrodes and urea electrodes, respectively.

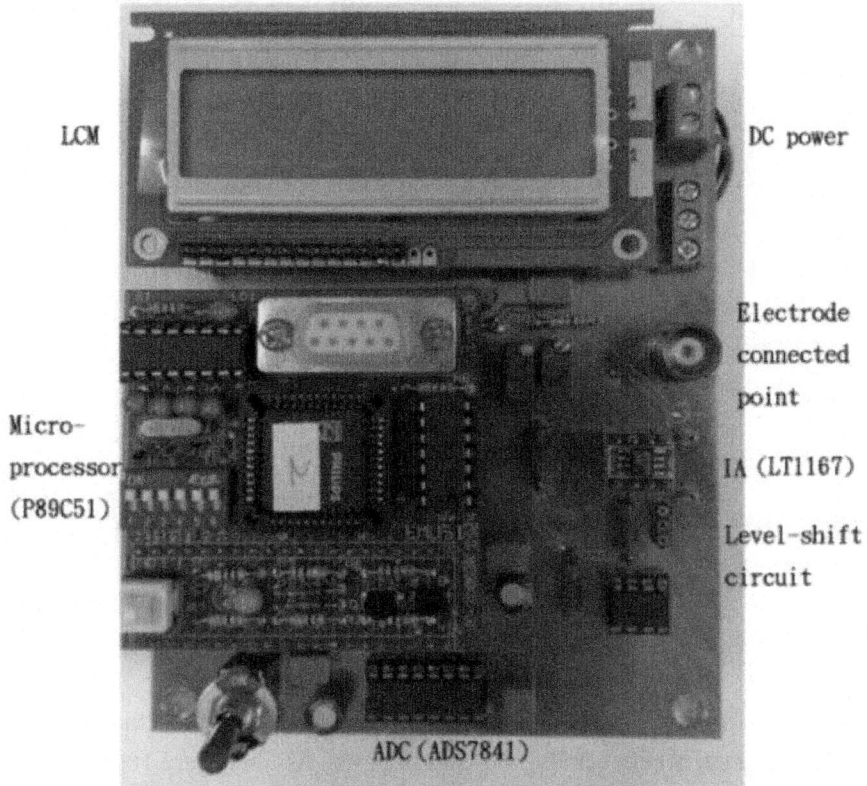

Figure 8: The designed electrochemical sensing instrumentation.

To achieve better accuracy, three-point calibration was adopted for our experimental measurement. The buffer solutions of pH 7, 10, and 4 were used for calibration based on interpolation and extrapolation techniques. From the measured voltages for the buffer solutions of pH 7 and 10, we obtained the pH sensitivity of 53.8 mV/pH. Similarly, from the measured voltages for the buffer solutions of pH 7 and 4, we obtained the pH sensitivity of 58.7 mV/pH. These two values were inputted into the designed program in P89C51. Using the calibrated system to measure the pH values of different solutions, the result is shown in Figure 9. The circle symbol is obtained by using commercially available pH meter (MP-512) for concentration measurement of hydrogen ion. The square symbol is obtained using our designed instrument. The measurement results in Figure 9 confirm the feasibility of the designed system. The obtained standard deviations using our instrument and MP-512 are 4.16 and 1.56, respectively. The three-point calibration method can be further extended for multi-point calibration to achieve higher accuracy.

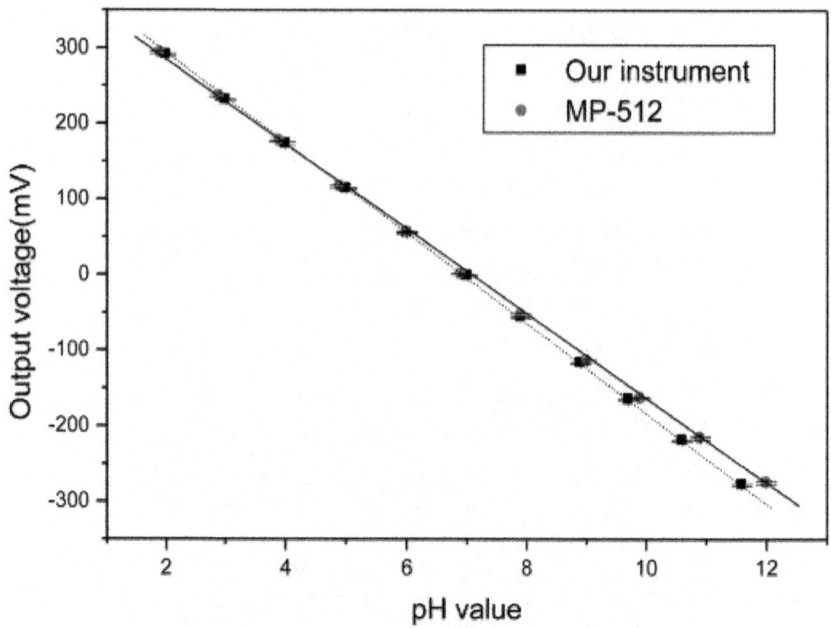

Figure 9: The plot of output voltage for different pH solutions.

The calibration curves of urea detection are shown in Figure 10. The dynamic range is between 0.25–64 mg/dL, as shown in Figure 10(a). In Figure 10(b), it can be observed that the linear range is between 0.25–2 mg/dL.

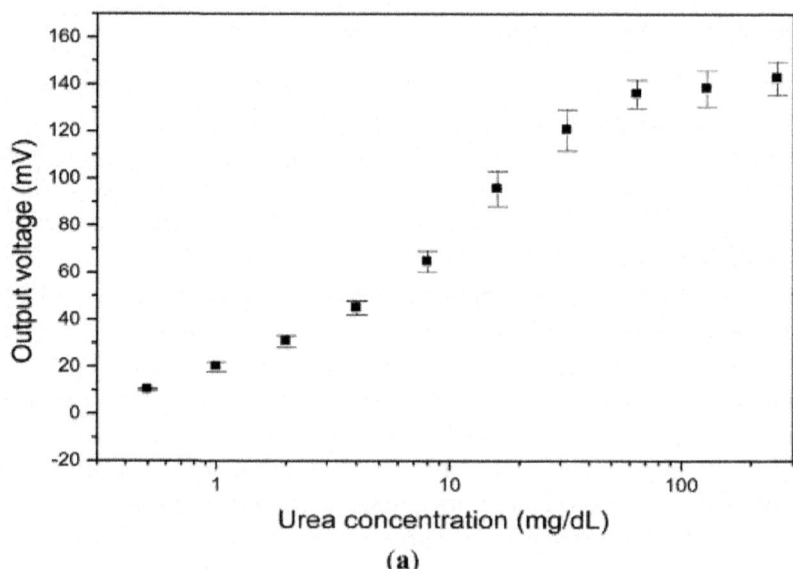

(a)

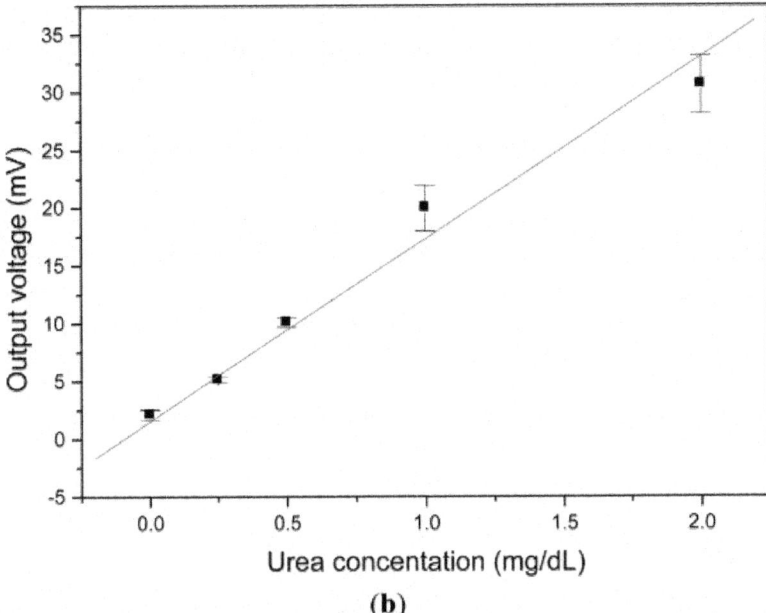

(b)

Figure 10: The output voltage for urea solutions with different concentrations. (**a**) Log scale; and (**b**) Linear scale.

The square of the sample correlation coefficient is 0.97 and the standard deviation is 1.82. Although there seems to be few points in the linear range, the non-linear problem can be solved using interpolation to calculate the measured voltages with microprocessor chip. By the blank determination method [12], the average output value of 1.2 mV and standard deviation of 0.6 mV were obtained for the blank, as shown in Figure 10(b). The limit-of-detection (LOD) and limit-of-quantification (LOQ) in this study are 0.25 mg/dL and 0.5 mg/dL, respectively. The concentration of urea nitrogen of 1 mg/dL is equivalent to a concentration of urea of 2.14 mg/dL. For a normal person, the concentration of urea nitrogen is in the range of 6 mg/dL to 24 mg/dL. Thus the system with measured concentration range of the urea in Figure 10 can be used for basic biomedical examinations.

CONCLUSIONS

An electrochemical potential measurement instrument using a potentiometric method is designed in this paper. It is applied to the concentration measurements of pH value and urea by connecting hydrogen ion-selective electrodes and urea electrodes. The urea concentration is obtained by virtue of the measurement of hydrogen ion concentration of solution after the reaction of urea sensing electrode and urea solution. Its feasibility and accuracy are considered by practical measurements. In the same way, with different sensing electrodes, the instrument can be used for the electrochemical measurement of different objects by slight adjustment of internal program. The instrument has the advantageous properties of multi-purpose, easy calibration and low cost. It could be further improved on volume, such as the integrated realizations of IA and ADC or IA, ADC and P89C51 microprocessor, since the operational amplifiers are the common core components for IA and ADC and these analog devices could be implemented in a single chip [13].

ACKNOWLEDGMENTS

This work has been supported by the National Science Council, Taiwan (Grant Nos NSC 101-2221-E-151-074 and NSC 100-2221-E-151-065). Many thanks are due to the reviewers for their useful comments.

REFERENCES

1.   Clark, L.C.; Lyons, C. Electrode system for continuous monitoring in cardiovascular surgery. *Ann. N.Y. Acad. Sci.* 1962,*102*, 29–45.

2.   Diamond, D. *Principles of Chemical and Biological Sensors*; Wiley: New York, NY, USA, 1998.

3.   Updike, S.J.; Hicks, G.P. The enzyme electrode. *Nature* 1967, *214*, 986–988.

4.   Bergveld, P. Thirty years of ISFETOLOGY: What happened in the past 30 years and what may happen in the next 30 years. *Sens. Actuators B Chem.* 2003, *88*, 1–20.

5.   Chi, L.L.; Chou, J.C.; Chung, W.Y.; Sun, T.P.; Hsiung, S.K. Study on extended gate field effect transistor with tin oxide sensing membrane. *Mater. Chem. Phys.* 2000, *63*, 19–23.

6.   Webster, J.G. *Medical Instrumentation Application and Design*; Wiley: New York, NY, USA, 2010.

7.   Wang, W.S.; Huang, H.Y.; Chen, S.C.; Ho, K.C.; Lin, C.Y.; Chou, T.C.; Hu, C.H.; Wang, W.F.; Wu, C.F.; Luo, C.H. Real-time telemetry system for amperometric and potentiometric electrochemical sensors. *Sensors* 2011, *11*, 8593–8610.

8.   Iniewski, K. *VLSI Circuits for Biomedical Applications*; Artech House Inc.: Norwood, MA, USA, 2008.

9.   Lin, C.Y.; Lai, Y.H.; Balamurugan, A.; Vittal, R.; Lin, C.W.; Ho, K.C. Electrode modified with a composite film of ZnO nanorods and Ag

nanoparticles as a sensor for hydrogen peroxide. *Talanta* 2010, *82*, 340–347.

10. Yin, L.T.; Chou, J.C.; Chung, W.Y.; Sun, T.P.; Hsiung, S.K. Separative structure ISFETs on a glass substrate. *Proc. SPIE* 1999, *3987*, 543–551.

11. Pan, C.W.; Chou, J.C.; Sun, T.P.; Hsiung, S.K. Development of the tin oxide pH electrode by the sputtering method.*Sens. Actuators B Chem.* 2005, *108*, 863–869.

12. Sanagi, M.M.; Ling, S.L.; Nasir, Z.; Hermawan, D.; Ibrahim, W.A.; Abu Naim, A. Comparison of signal-to-noise, blank determination, and linear regression methods for the estimation of detection and quantification limits for volatile organic compounds by gas chromatography. *J. AOAC Int.* 2009, *92*, 1833–1838.

13. New 8FX MB95430H Series Datasheet. Available online: http://www. fujitsu.com/downloads/MICRO/fma/mcu/MB95430H_DS_Mar2011. pdf (accessed on 4 April 2011).

Chapter 9

EVALUATION OF THREE INSTRUMENTATION TECHNIQUES AT THE PRECISION OF APICAL STOP AND APICAL SEALING OF OBTURATION

Özgür Genç[I], Tayfun Alaçam[II], Guven Kayaoglu[II]

[I]DDS, PhD, Department of Restorative Dentistry and Endodontics, Faculty of Dentistry, Yüzüncü Yıl University, Van, Turkey

[II]DDS, PhD, Department of Restorative Dentistry and Endodontics, Faculty of Dentistry, Gazi University, Ankara, Turkey

ABSTRACT

Objective: The aim of this study was to investigate the ability of two NiTi rotary apical preparation techniques used with an electronic apex locator-integrated endodontic motor and a manual technique to create an apical stop at a predetermined level (0.5 mm short of the apical foramen) in teeth with disrupted apical constriction, and to evaluate microleakage following obturation in such prepared teeth.

Material and Methods: 85 intact human mandibular permanent incisors with single root canal were accessed and the apical constriction was disrupted using a #25 K-file. The teeth were embedded in alginate and instrumented to #40 using rotary Lightspeed or S-Apex techniques or stainless-steel K-files. Distance between the apical foramen and the created apical stop was measured to an accuracy of 0.01 mm. In another set of instrumented teeth, root canals were obturated using gutta-percha and sealer, and leakage was tested at 1 week and 3 months using a fluid filtration device.

Results: All techniques performed slightly short of the predetermined level. Closest preparation to the predetermined level was with the manual technique and the farthest was with S-Apex. A significant difference was found between the performances of these two techniques ($p < 0.05$). Lightspeed ranked in between. Leakage was similar for all techniques at either period. However, all groups leaked significantly more at 3 months compared to 1 week ($p < 0.05$).

Conclusions: Despite statistically significant differences found among the techniques, deviations from the predetermined level were small and clinically acceptable for all techniques. Leakage following obturation was comparable in all groups.

INTRODUCTION

Endodontic treatment should be performed within the confines of the root canal, avoiding overextension of instruments and filling materials into the periradicular tissues. The most favorable healing is achieved when these principles are complied with, and extrusion of the filling material has been found to be associated with severe inflammatory reaction[12].

Apical constriction is considered to be the part of the root canal with the smallest diameter and is located 0.5-1.5 mm inside the apical foramen[16]. It is generally accepted as the terminal point the endodontic treatment should extend to. The preservation of the apical constriction helps maintenance of the endodontic instruments, chemicals and filling materials within the root canal. However, this anatomical structure may not have formed in immature teeth or may be lost iatrogenically (e.g., instrumentation beyond the apical foramen), surgically (e.g., apical resection) or due to an inflammatory resorption. Then, in such cases, a new apical stop (apical seat, apical matrix, apical ledge) should be created to confine the instruments and chemicals to the canal and to supply a barrier against which gutta-percha can be compacted[3,6]. This apical stop is prepared through the use of successively larger files to the working length. While apical stop preparation can be done using conventional stainless-steel files, rotary NiTi files produced especially for the instrumentation of the apical part of the root canal can also be used for this purpose.

Besides proper instrumentation of the root canal, another goal of endodontic treatment is to hermetically obturate the root canal system. In this case, ingress of tissue fluids and microorganisms into the canal space and any possible microbial activity within the canal could be eliminated or at least reduced[13,20]. It is possible that instrumentation-related factors such as the shape of the apical preparation achieved using different instrumentation systems or the debris remaining on the dentinal walls following instrumentation could affect the adaptation of the root canal filling material to canal walls and thus have an impact on microleakage.

The ability of apical preparation instruments for preparing an apical stop at predetermined levels and the microleakage occurring following obturation of such prepared root canals has not been investigated so far. This study aimed to test the ability of two NiTi rotary apical preparation techniques used with an apex locator-integrated endodontic motor, and a stainless-steel manual technique to create a new apical stop at a predetermined level in teeth with disrupted apical constriction, and to assess microleakage following obturation.

MATERIAL AND METHODS

Selection and Preparation of the Teeth

Eighty-five extracted intact human mandibular permanent incisors were selected after examination with stereomicroscopy and digital radiography. All the teeth were free of cracks, had single root canals and mature apices. The root surfaces were scaled with a periodontal curette to remove the tissue debris and the teeth were stored in physiologic saline solution.

Endodontic access cavities were opened using high-speed burs under water spray. The pulp was extirpated and a #25 K-file was inserted into the root canal and extruded 1 mm past the apical foramen. The coronal two-thirds were enlarged using #1 and #2 Gates-Glidden burs. The teeth were embedded in alginate in plastic molds.

Instrumentation Techniques

Two rotary NiTi apical preparation techniques and a manual stainless-steel preparation technique were tested. The rotary NiTi apical preparation techniques included: Lightspeed (Lightspeed Technology Inc., San Antonio, TX, USA) and S-Apex (FKG Dentaire, La Chaux-de-Fonds, Switzerland). Tri Auto ZX (J Morita Co., Kyoto, Japan), an electronic apex locator-integrated endodontic motor was used along with the NiTi techniques. The device was adjusted to the H (high torque) mode and set to 0.5 mm (distance short of the apical foramen) and used at auto apical reverse function. The lip clip was inserted into the alginate. The manual instrumentation group employed stainless-steel K-files (Dentsply/DeTrey, Konstanz, Germany) used with a standardized technique described previously[15]. Briefly, the file was used first with a quarter clockwise rotation followed by a pull-back motion and used repeatedly until it became loose at the working length. Before instrumenting in the manual group, working lengths were determined using Tri Auto ZX at the apex locator mode (EMR mode) and the device set to 0.5 mm. The rubber stops of the hand-files were adjusted to the registered lengths. Instrumentation was from #30 to #40 in all groups.

A #25 K-file was used to maintain apical patency after each instrument. The root canals were irrigated with 10 mL of 5.25% NaOCl between each instrument change. Final irrigation was with 10 mL of 17% EDTA for 60 sec, followed by 10 mL of NaOCl irrigation.

Measurement of the Level of the Created Apical Stop

Thirty teeth instrumented as described (n=10 for each instrumentation technique) were allocated for testing the ability of each instrumentation technique to approximate to the predetermined level (0.5 mm short of the apical foramen). Once the apical preparation was accomplished, the final instrument was reinserted into the root canal without excessive pressure and advanced to the level it reached freely, then fixed in place with a flowable composite resin. The teeth were taken out of the alginate molds. An ohmmeter (YFE, Hsin-Chu City, Taiwan, R.O.C.) was used in order to measure the distance between the apical foramen and the level of the created apical stop. Briefly, one pole of the ohmmeter was connected to the shank of the fixed instrument in the canal, and the other pole was connected to the shank of a #15 finger spreader (Figure 1). The spreader was inserted retrogradely from the apical foramen through the root canal and advanced until the display of the ohmmeter indicated a contact. The distance the spreader advanced until the contact was registered by marking on the spreader with an acetate marker. The distance between the tip of the spreader and the mark was measured under stereomicroscope using a digital caliper to an accuracy of 0.01 mm and this value was considered the distance between the apical foramen and the apical stop.

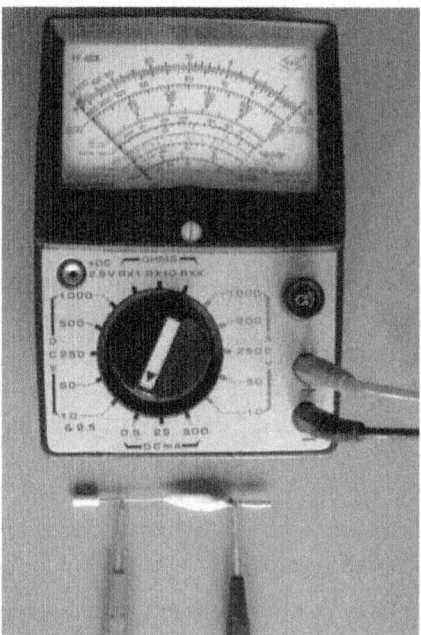

Figure 1: The experimental setting showing the ohmmeter with the poles connected to the files.

The measurements were undertaken by two investigators. No significant difference was found between the measurements of these investigators when paired*t*-test was performed (p=0.90). The average of the two measurements for each sample was calculated and the data were analyzed statistically using the Anderson-Darling normality test and the Tamhane test at p=5% as the level of significance.

Endodontic Filling and the Leakage Test

Forty-five teeth instrumented as described (n=15 for each instrumentation technique) were allocated for testing the microleakage after root-filling of the instrumented teeth. The teeth were taken out of the alginate molds. The root canals were dried using paper points and filled using gutta-percha and sealer (AH Plus, Dentsply/DeTrey, Konstanz, Germany) with the lateral condensation technique. A size 40 gutta-percha cone was used as the master cone. The gutta-percha cones were coated with sealer before insertion into the canal. The teeth were checked radiographically (RVG, Trophy Radiologie, Paris, France) from the labial and lateral aspects for the quality of the root filling (e.g. homogeneity of the filling or presence of a void). All root surfaces were covered with nail polish with care to avoid the apical foramen. The access cavities were sealed with a temporary filling. Ten endodontically accessed-teeth were not obturated and served as positive control. The teeth were stored in physiological saline with 0.2% sodium azide solution until and between the leakage experiments.

The root fillings were tested for leakage using a fluid filtration assembly described elsewhere7. In this method, a pressure of 2 psi (0.136 atm) was applied with O2 to force water through voids within a filled canal. The fluid flow was quantified by observing the movement of an air bubble created within a 25 µL micropipette. This micropipette located between the pressure source and the sample to be tested. Measurements of the fluid movement were made at 2-min intervals for 8 min and averaged. The values were expressed as µL/min/cmH2O. The experiment was performed at 1-week and 3-month post-obturation periods. The normality of the data was analyzed using the Anderson-Darling test. The differences at each period between the groups were statistically evaluated using the Tamhane test. The difference for each group between the 1-week and 3-month measurements was analyzed statistically using the paired samples t-test at p=5% as the level of significance.

RESULTS

Approximation to the Predetermined Level

The distance from the apical foramen for each instrumentation technique was as follows [average distance mm(SD mm)]: manual technique [0.59(0.11)], LightSpeed [0.91(0.45)] and S-Apex [1.18(0.29)]. Thus, average proximity to the predetermined level was: manual technique: 0.09 mm; Lightspeed: 0.41 mm and S-Apex: 0.68 mm. There was a significant difference between the manual technique and S-Apex ($p < 0.05$). No significant difference was found between Lightspeed and the other techniques (Figure 2).

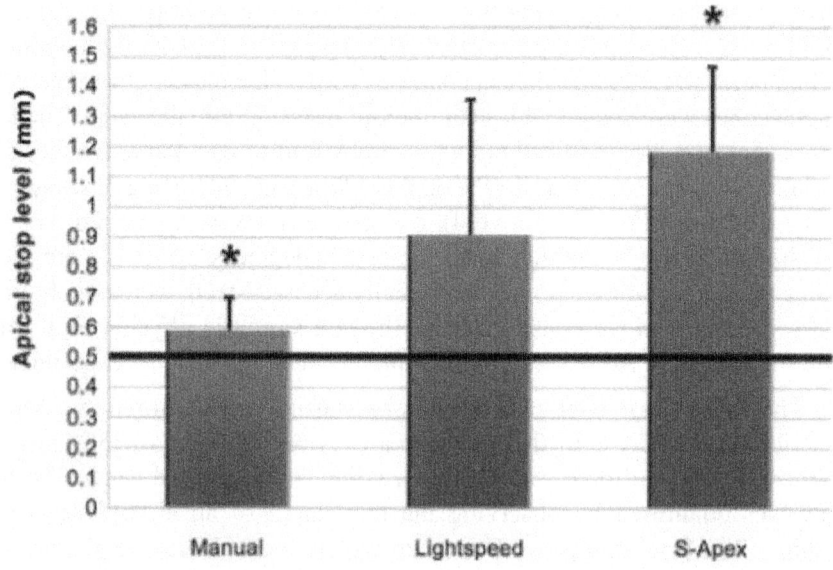

Figure 2: Graph showing the approximation by the three instrumentation techniques to the predetermined level (means and standard deviations in millimeters; n=10). 0 is the apical foramen and 0.5 is the predetermined termination level. The asterisk (*) denotes the groups between which a statistically significant difference exists.

Leakage Following Obturation

The experimental results are shown in Table 1. No significant difference was found between the techniques at either 1-week or 3-month measurements ($p > 0.05$). However, the positive control group leaked significantly more than the other groups at either period ($p < 0.05$).

Table 1: Fluid filtration measurements for the experimental groups at 1-week and 3-week and 3-month peroids

Instrumentation technique	1-week, Mean (±SD) leakage (µL/min/cmH$_2$O)	3-month, Mean (±SD) leakage (µL/min/cmH$_2$O)
Lightspeed	1.02x10^{-3} (3.22x10^{-4})^{Aa}	2.40x10^{-3} (1.24x10^{-3})^{Ab}
S-Apex	9.22x10^{-4} (4.15x10^{-4})^{Aa}	1.12x10^{-3} (4.46x10^{-4})^{Ab}
Manual	1.09x10^{-3} (4.74x10^{-4})^{Aa}	1.31x10^{-3} (5.43x10^{-4})^{Ab}
Positive control	78.62 (4.75x10^{-1})^{Ba}	78.95 (1.00)^{Ba}

The capital letters (A,B) indicate statistical difference between the groups within either 1-week or 3-month periods. Groups with the same capital letter are not significantly different (p>0.05). SD=standard deviation
The lower case letters (a,b) indicate statistical difference between the 1-week and 3-month measurements for each group. Groups with the same lower case letter are not significantly different at the two measurement periods (p>0.05).

Leakage at 3 months was significantly greater for all techniques compared to 1-week (p<0.05). No significant difference was found between the leakage at 1-week and 3-month for the positive control group (p>0.05).

DISCUSSION

Rotary NiTi instrumentation has gained popularity in endodontics in recent years. While preparation of the root canal in conventional technique is done using single type of instrument (e.g. successive size K-files), rotary NiTi instrumentation employs different and specific instruments for the preparation of the coronal, mid and apical thirds of the canal. Lightspeed and S-Apex are examples to the instruments designed for the apical third. This study tested how close these novel systems and a conventional manual technique would approximate to a predetermined level in teeth with disrupted apical constriction with a purpose to create a new apical stop. The effect to microleakage of instrumentation with these techniques was also evaluated.

The major findings of this study were that all instrumentation techniques prepared slightly short of the predetermined level; the manual technique prepared closest, and the S-Apex technique prepared farthest to the predetermined level. The Lightspeed technique ranked in between. However, in spite of these variations, leakage following obturation was comparable in all groups.

The NiTi rotary instruments in this study were used along with Tri Auto ZX, an electronic apex locator-integrated endodontic motor. Previous studies employing Tri Auto ZX at a setting of 0.5 found that measurements using hand-files or rotary NiTi files were slightly short of the predetermined level[4,9]. The findings of the study presented here are in line with these studies.

The finding that approximation with S-Apex to the predetermined level was less satisfactory than the other techniques can be explained by some factors. The most important factor is that the S-Apex files were not used as per the manufacturer's recommendations: while a rotational speed of 500-1000 rpm has been recommended by the manufacturer for the optimum operation of the S-Apex files[2], the Tri Auto ZX device provides a constant speed of 280±50 rpm which cannot be changed[5]. Rotation at slow speed may be the main reason of the inferior performance of the S-apex system. Another factor is the shape of the cutting portion of the instrument. Different from the others, S-Apex has an inverted cone cutting portion. In the gradually narrowing canal of the mandibular incisor, this shape could have disabled the progress of the file to the predetermined level. The manual instrumentation in this study gave the best result. One explanation for this may be that the manual control of a file was simpler than the control of a rotary file mounted on a handpiece. It was "controlled-preparation" to the initially established working length in the manual group. However, the rotary groups followed a more sophisticated technique: it was a dynamic process including the simultaneous measurement of the working length and rotation of the files. A recent study also reported inequivalent apical measurements obtained with motor-driven apex locators (using NiTi files) and their manual apex-locating modes (using stainless-steel files)[1]. However, we believe that the difference is not due to the alloys the files used in these techniques were made of. In two studies, for example, no significant difference was found between the electric length measurements achieved by the use of stainless-steel or NiTi rotary instruments in the same root canal[10,14]. Regarding these, the differences found in this study between the performances of the different instrumentation systems may be attributed to their designs, handling and to technical shortcomings. Nevertheless, all the instrumentation techniques tested in this study, with small deviations from the predetermined level, are considered to be clinically acceptable. It has been suggested that best treatment results in cases with necrotic pulp were achieved when treatment procedures were terminated within 2 mm (0 to 2 mm) of the radiographic apex[18]. The results of the study presented here indicated that the tested techniques could perform within this range.

In spite of the varying deviations from the predetermined level, leakage was similar in all groups. This may be because there were small (in a millimeter scale only) differences between the groups regarding approximation to the predetermined level. Another explanation is that the results of the fluid filtration experiment reflect the seal of the root canal filling as whole. Even if increased leakage existed at the apical part, this could have been blockaded by the filling at the middle or coronal parts of the canal, and thus the fluid flow in the experimental assembly could have been impeded.

This study further found that the leakage increased over time. Leakage in all groups was significantly greater in the 3-month measurements than in the 1-week measurements. This is probably due to disintegration of the sealer during storage. The finding that the leakage in obturated root canals increased with time is in agreement with previous studies where zinc oxide-eugenol-based and epoxy resin-based sealers were used[8,11,17,19].

CONCLUSIONS

In conclusion, there were significant but minor differences (less than a millimeter) among the instrumentation techniques regarding approximation to 0.5 mm short of the apical foramen as the predetermined level. However, despite these differences, leakage was similar in all groups following obturation of the root canals.

ACKNOWLEDGEMENT

This study has constituted part of the doctorate thesis by Özgür Genç entitled «Approximation of different apical files to predetermined level in overinstrumented roots and effects on the apical seal», Gazi University, Ankara, 2007.

This study was financially supported by the Gazi University Research Projects Fund.

REFERENCES

1.   Barthelemy J, Gregor L, Krejci I, Wataha J, Bouillaguet S. Accuracy of electronic apex locater-controlled handpieces. Oral Surg Oral Med Oral Pathol Oral Radiol Endod. 2009;107:437-41.

2.   FKG Dentaire. NiTi S-Apex [online]. La Chaux-de-Fonds. [cited 27 Dec. 2010]. Available from: <http://www.fkg.ch/fileadmin/template/main/images/download/datasheets/fkg_datasheet_sapex_an_lowr.pdf>   .

3.   Gomes-Filho JE, Hopp RN, Bernabé PF, Nery MJ, Otoboni JA Filho , Dezan E Júnior. Evaluation of the apical infiltration after root canal disruption and obturation. J Appl Oral Sci. 2008;16:345-9.

4.   Grimberg F, Banegas G, Chiacchio L, Zmener O. *In vivo* determination of root canal length: a preliminary report using the Tri Auto ZX apex-locating handpiece. Int Endod J. 2002;35:590-3.

5.   J Morita. Tri Auto ZX [online]. [cited 27 Dec. 2010]. Available from: <http://www.morita.com/usa/root/img/pool/pdf/product_brochures/tri_auto_ps_1-227_1008.pdf>   .

6.   Kast'áková A, Wu MK, Wesselink PR. An *in vitro* experiment on the effect of an attempt to create an apical matrix during root canal preparation on

coronal leakage and material extrusion. Oral Surg Oral Med Oral Pathol Oral Radiol Endod. 2001;91:462-7.

7. Kont Cobankara F, Adanir N, Belli S, Pashley DH. A quantitative evaluation of apical leakage of four root-canal sealers. Int Endod J. 2002;35:979-84.

8. Kopper PM, Vanni JR, Della Bona A, Figueiredo JA, Porto S. *In vivo* evaluation of the sealing ability of two endodontic sealers in root canals exposed to the oral environment for 45 and 90 days. J Appl Oral Sci. 2006;14:43-8.

9. Mente J, Seidel J, Buchalla W, Koch MJ. Electronic determination of root canal length in primary teeth with and without root resorption. Int Endod J. 2002;35:447-52.

10. Nekoofar MH, Sadeghi K, Sadighi Akha E, Namazikhah MS. The accuracy of the Neosono Ultima EZ apex locator using files of different alloys: an *in vitro* study. J Calif Dent Assoc. 2002;30:681-4.

11. Pommel L, Camps J. *In vitro* apical leakage of system B compared with other filling techniques. J Endod. 2001;27:449-51.

12. Ricucci D, Langeland K. Apical limit of root canal instrumentation and obturation, part 2. A histological study. Int Endod J. 1998;31:394-409.

13. Sedgley CM. The influence of root canal sealer on extended intracanal survival of *Enterococcus faecalis* with and without gelatinase production ability in obturated root canals. J Endod. 2007;33:561-6.

14. Thomas AS, Hartwell GR, Moon PC. The accuracy of the Root ZX electronic apex locator using stainless-steel and nickel-titanium files. J Endod. 2003;29:662-3.

15. Tinaz AC, Alacam T, Uzun O, Maden M, Kayaoglu G. The effect of disruption of apical constriction on periapical extrusion. J Endod. 2005;31:533-5.

16. Vertucci FJ. Root canal morphology and its relationship to endodontic procedures. Endod Top. 2005;10:3-29.

17. Von Fraunhofer JA, Fagundes DK, McDonald NJ, Dumsha TC. The effect of root canal preparation on microleakage within endodontically treated teeth: an *in vitro* study. Int Endod J. 2000;33:355-60.

18. Wu MK, Wesselink PR, Walton RE. Apical terminus location of root canal treatment procedures. Oral Surg Oral Med Oral Pathol Oral Radiol Endod. 2000;89:99-103.

19. Xu Q, Ling J, Cheung GS, Hu Y. A quantitative evaluation of sealing ability of 4 obturation techniques by using a glucose leakage test. Oral Surg Oral Med Oral Pathol Oral Radiol Endod. 2007;104:109-13.

20. Yücel AC, Ciftçi A. Effects of different root canal obturation techniques on bacterial penetration. Oral Surg Oral Med Oral Pathol Oral Radiol Endod. 2006;102:e88-92.

Chapter 10

TRUE UNIPOLAR ECG MACHINE FOR WILSON CENTRAL TERMINAL MEASUREMENTS

Gaetano D. Gargiulo

The MARCS Institute, University of Western Sydney, Kingswood, NSW 2747, Australia

ABSTRACT

Since its invention (more than 80 years ago), modern electrocardiography has employed a supposedly stable voltage reference (with little variation during the cardiac cycle) for half of the signals. This reference, known by the name of "Wilson Central Terminal" in honor of its inventor, is obtained by averaging the three active limb electrode voltages measured with respect to the return ground electrode. However, concerns have been raised by researchers about problems (biasing and misdiagnosis) associated with the ambiguous value and behavior of this reference voltage, which requires perfect and balanced contact of at least four electrodes to work properly. The Wilson Central Terminal has received scant research attention in the last few decades even though consideration of recent widespread medical practice (limb electrodes are repositioned closer to the torso for resting electrocardiography) has also sparkled concerns about the validity and diagnostic fitness of leads not referred to the Wilson Central Terminal. Using a true unipolar electrocardiography device capable of precisely measuring the Wilson Central Terminal, we show its unpredictable variability during the cardiac cycle and confirm that the integrity of cardinal leads is compromised as well as the Wilson Central Terminal when limb electrodes are placed close to the torso.

INTRODUCTION

Surface electrocardiography, by definition, is the timedomain representation of the electrical activity of the beating heart inside the chest, measured as voltage variation over time by surface electrodes placed in contact with the skin. Surface electrocardiography is represented by a vector quantity $(\bar{P})$

rotating around a fixed point (the electrical center of the heart) in the body frontal plane describing an angle (α) with a fixed direction identified by an imaginary line crossing the shoulders 1. This definition was originally outlined in 1908 by E. Einthoven, later revised in 1931 by F. N. Wilson, who named the fixed point as the "central terminal," and further modified in 1942 by E. Goldberger, who invented the augmented leads 1. From 1942, the mentioned definition and associated recording guidelines produced the so-called 12-lead ECG system, which is currently considered to be the best practice 1, 2. The 12-lead ECG is so called because it produces twelve ECG signals. It uses a reference electrode placed on the right leg (RL) and nine exploring electrodes: three limb electrodes placed on the right arm (RA), left arm (LA), and left leg (LL) and six electrodes placed over the torso near the heart 1. Electrode positioning and signals recordable from the six electrodes over the torso have been named precordial leads (precordials) and are also known simply as "chest leads" (see Figure 1(a)) or as V_1 to V_6 leads, while the signals recordable from the limbs have been named cardinal (or fundamental) Einthoven leads (see Figure 1(b)) and are referred to as Lead I, Lead II, and Lead III or simply as "limb leads":

Lead I: $V_I = \Phi_L - \Phi_R$;

Lead II: $V_{II} = \Phi_F - \Phi_R$;

Lead III: $V_{III} = \Phi_F - \Phi_L$;

   With

V_I being the voltage of Lead I;

V_{II} the voltage of Lead II;

V_{III} the voltage of Lead III;

Φ_L the potential at the left arm*;

Φ_R the potential at the right arm*;

Φ_F the potential at the left leg*

*referred to the electrode on the right leg (Φ_{RL}).

   The augmented leads are measured as the voltage difference between each of the limb potentials and the average of the other two limb potentials. For example, the augmented lead aV_F is measured as

$$aV_F = \Phi_F - \frac{(\Phi_L + \Phi_R)}{2}.$$

(1)

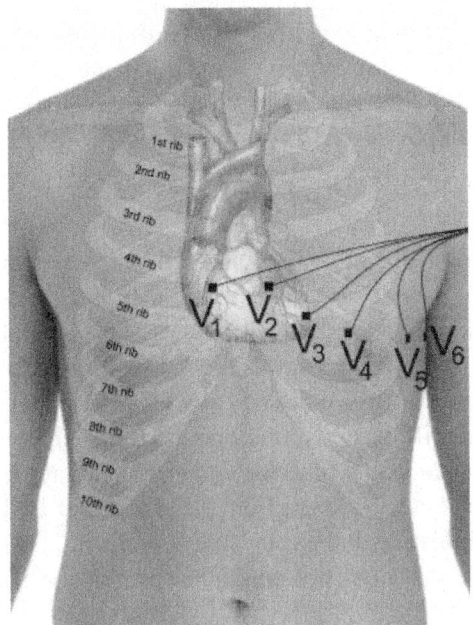

(a) Precordial placement

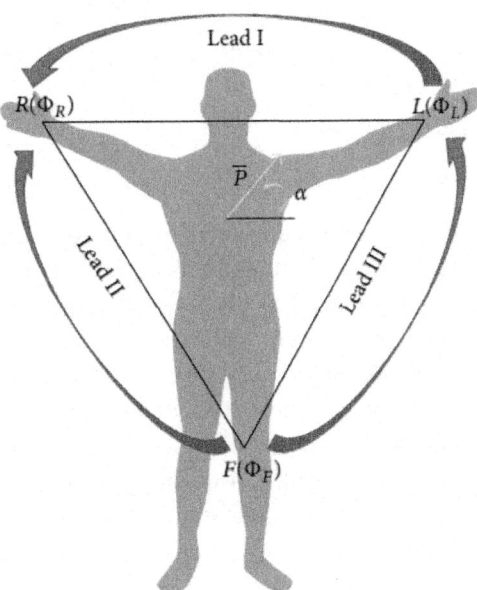

(b) Limb leads and electrodes

Figure 1: Twelve-lead ECG electrode placement and lead names 1. (a and b).

Because all of the limb potentials are implicitly referred to the potential of the right leg, it is possible to infer that cardinal leads are recorded as twice the voltage difference. For example, assuming the potential of the right leg Φ_{RL} being measured with respect to a point at a neutral potential (i.e., earth ground), Lead I can be rewritten as

$$V_I = \left(\Phi_L - \Phi_{RL}\right) - \left(\Phi_R - \Phi_{RL}\right).$$
(2)

Although at first glance it may seem that the potential Φ_{RL} will cancel out; due to the nonideal (not infinite) capacity to reject common signals that are present simultaneously at the inputs, known as the Common Mode Rejection Ratio (CMRR) 3, 4. of the employed amplifiers, any contact imbalance between the three electrodes may cause signal quality degradation and unpredictable drift of slow components. Intuitively, the effect of contact impedance imbalance gets worse when considering augmented leads as they require perfectly balanced contact from all of the four limb electrodes 1, 5, 6.. This is counterintuitive as the circuit that they form in the human body is an equilateral triangle that does not take into account the RL voltage at all (see Figure 1(b)).

Similarly, the voltage of a virtual point called the Wilson Central Terminal (WCT) is subtracted from each of the precordials' electrode potentials. The WCT is obtained by averaging the potential at the limbs referred to the reference electrode on the right leg using three identical resistors (5 kΩ or higher) connected to a single point 1.:

$$\Phi_{WCT} = \frac{\left(\Phi_L + \Phi_R + \Phi_F\right)}{3}.$$
(3)

Although Wilson himself used to refer to the precordials as "unipolar" 7., this has been repeatedly pointed out as a misnomer due to the repeated voltage difference required to obtain them 8–11.. It has also been demonstrated that the WCT cannot be considered a "null" potential 8, 9. nor should it be confused with the real center of the heart potential, because the ECG signals travel through different trunks of an inhomogeneous volume conductor and can be exposed to different sources of noise such as different expositions to RF fields and artifacts 9, 12.. In 1954, Frank 8. was the first to raise concerns about the potential fluctuations in the WCT during a cardiac cycle and how they could bias the ECG measurement

8, 13, 14.. He predicted that within a few years a new, refined cardiac conduction theory and ECG system able to work without the WCT would emerge. In the early days of modern electrocardiography, other researchers were also able to confirm that the WCT is not constant during the cardiac cycle.

Confirmation of errors and variability of the WCT during the cardiac cycle have been measured employing an "integrator electrode." This procedure requires the entire human body to be encased in a metal structure and then immersed in water (neutral reference) during the measurement of ECG. Unfortunately, due to the cumbersomeness of the measurement process, this technique was used only for few experimental trials 15,16. In recent years, the significance of theWCT and even its physical location has also been debated 9,10,17. However, aside from notable attempts in the 1940s and 1950s 14, 18, 19, until our study, the WCT has never been correctly measured without a cumbersome procedure and in a repeatable way.

In this context, one must mention that not only has the WCT received scant research attention in the last few decades, but also there is a generalized lack of modern studies about the general placement of electrodes and the impact that electrode misplacement (particularly when intentional) may have upon diagnosis. Current common widespread medical practice is to move the limb electrodes to positions closer to the torso (shoulders and hips or sides of the navel). This is thought to reduce the obtrusiveness of the ECG recording as cables are not spread all over the body, which is particularly advantageous during stress recordings. However, there is evidence 20 that limb electrode positioning that affects the QRS influences the diagnosis of ischemic (including chronic) heart diseases 21, 22.. Although there is some evidence that in healthy subjects the variation in the ECGs imposed by alteration of the limb electrodes can be classified only as statistically relevant and not as clinically relevant 23., due to the significant shift in cardiac axis and waveform amplitude that can be observed in both ECG planes when the limb electrodes are in positions different from the standard ones 24., standardized recommendation for ECG clinical practice 25. confirms that misplacement of limb electrodes should be avoided 22 or used only where strictly necessary (i.e., stress test) and always noted on the recording 25..Over the past two years, we have developed a new electrocardiographic device 3, 11, 12, 26–28 that allows realtime visualization and precise measurement of the WCT amplitude, shape, and variations; using this device we show that the WCT exhibits a clinically significant variation (>0.1 mV or >1 mm 2, 14.) across different recordings and during the course of the same recording. For the evaluation presented in this paper we have partially reused the unipolar ECG data that have been recorded from a small population of healthy subjects who volunteered during a previous study 11, 12, 26, 27. and agreed to have the data analyzed for publication purposes by expert cardiologists. The subject population comprises five males covering the age span of 29–36 years with an average age of 32.5 years. None of the subjects had a history of cardiac illness, and all the recordings presented normal sinus rhythms. We also recorded data from one volunteer subject again, performing two recordings consecutively to show the effect of placing the limb electrodes near the torso on cardinal leads.

EXPERIMENTAL SECTION

Our principal hypotheses for this study are as follows.

(1)   The WCT is not a stable voltage reference exhibiting a clinically significant voltage variation.

(2)   Moving the limb electrodes to a position near the torso can affect the shape and amplitude of cardinal leads as well as the WCT.

To demonstrate our hypotheses, we firstly introduce the true unipolar machine and a measurement technique that allows us to reliably measure and store the WCT; then, we present the data processing with a full example of WCT variability across the cardiac cycle and through a recording. Lastly, we show the effect that the placement of the limb electrodes near the torso (from ankles and wrists to hips, sides of the navel, and shoulders) has on limb leads and the WCT 25.

Hardware Development

Our hardware front-end and its pilot evaluation are properly described in 11, 12, 26– 28. However, for the sake of completeness, in this section we give a brief summary of the measurement hardware employed in this study. In Figure 2, we show a functional block diagram of the ECG amplifier (one single channel). In principle, we regard the unipolar ECG measurement as a combined observation of noise and useful signal. It is thus possible to measure the local signal of interest by subtracting the local noise (or what is regarded as such) from the measured signal. As it is possible to observe in Figure 2, the measured signal (measurement electrode) is fed to an instrumentation amplifier that subtracts from the signal a low-pass version of the same signal (the low-pass cut-off frequency is set at 0.1 Hz). With this technique, a pseudo-high-pass DC-coupled ECG front-end is achieved, preserving the ultrahigh input of the amplifier, which allows the use of dry electrodes. Experiments confirmed that the low-pass filter used to achieve the pseudo-high-pass filter can be implemented with passive components and its cutoff frequency can be positioned at very low frequency (i.e., 0.01 Hz), employing high value capacitors and resistors. This is possible because the ultrahigh-input impedance of the instrumentation amplifier employed can cope with several MΩ of impedance.

Amplifier referencing is achieved via the reference terminal of the instrumentation amplifier labeled as "Ref." The Ref terminal receives a damped version (low passed) of the summation of all the electrode signals and of the RL electrode. This technique, which is also known as "modified ground bootstrapping" 3, 12, 29–31., similar to the standard ground bootstrapping 3, 32., achieves power-line noise and electrodic noise suppression without the use of a driven rightleg technique 33, 34.

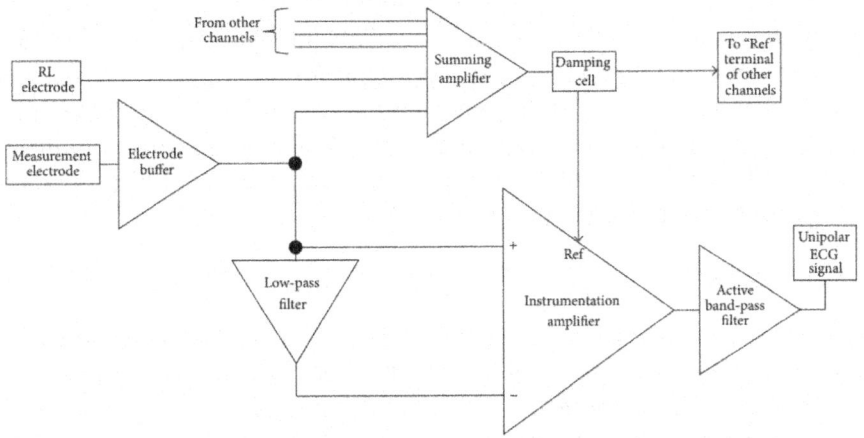

Figure 2: Block diagram of proposed ECG system.

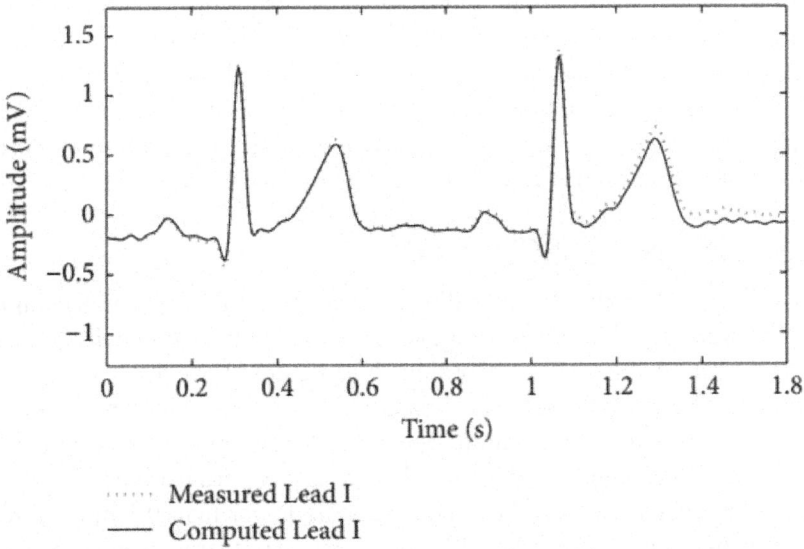

Figure 3: Example of a traditional ECG lead reconstruction from unipolar leads (point-to-point subtraction) (the data used to plot the image were recorded for the study 12.).

Signals recorded using this instrument can be regarded as being referred directly to the right leg. Therefore, a simple point-by-point subtraction between recorded signals allows real-time calculation of the 12-lead ECG. In Figure 3, an example of the calculation for Lead I is shown. In this example, prerecorded left-arm and right-arm signals have been simply subtracted to obtain Lead I. With this recording technique, the WCT is simply calculable from a point-bypoint average of the recorded limb potentials. In order to allow reconstruction of traditional precordials (obtained by simple point-by-point subtraction of the WCT), our precordials are also directly referred to the potential of RL 11,12, 27.. In our previous pilot study 12, 28., we demonstrated that correlation between the reconstructed signals and parallel recording of traditional signals exceeds 90% with minimal differences, which are due to components' tolerance 11, 12, 26..

Measurement

For this study, we calculate the WCT by averaging the prerecorded limb potentials. As we have shown in our previous analysis, the WCT is profoundly different across subjects and may have the shape of ECG leads with sometimes very well marked characteristic waveforms such as a P wave, a QRS complex, and a T wave. For this reason, we measure the WCT's amplitude at its largest feature that is expected to normally coincide with the QRS-like complex. In other words, we measure this amplitude as the peak-topeak amplitude. In this study, we show that the amplitude of the WCT varies during a recording and that, similar to what has been already demonstrated for standard ECG leads 20., its shape and amplitude are affected by the positions of limb electrodes. Using a case study we have also been able to justify the commonly observed shift of the cardiac axis towards the vertical direction 20, 23, 24.

RESULTS AND DISCUSSION

(1)   The WCT exhibits clinically relevant (>0.1 mV or >1 mm) amplitude variability during each cardiac cycle as well as clinically significant variation during the recording. In order to show this variability in a concise way, we selected a random starting point within the recording and measured the amplitude of the WCT for 10 consecutive beats after that point. As it is possible to observe from Figure 4, all of the 10 considered beats have an amplitude larger than 0.1 mV; moreover, between beat #3 and beat #6 there is the largest large extent of variability (0.12 mV) between cardiac cycles.

(2)   Similar analyses performed for the other subjects of our database 11, 12, 26, 27. yield similar results.

(3)   Our general WCT amplitudes are in accordance with values presented
       in the literature. We recall that amplitudes for the WCT of the order of
       0.2 mV were already measured during a historical experiment that made
       use of a cumbersome procedure. During the experiment a volunteer
       was immersed in water whilst being encased in a metal structure called
       an "integrator electrode" 15, 16, 18, 35.. Our device instead allows
       continuous WCT precise measurement by recording straight from the
       limb electrodes.

(4)   The WCT noise level is influenced directly by all three limb potentials;
       hence movement artifacts on any of the limbs or any contact impedance
       imbalance between the limb electrodes will directly affect the WCT
       signal quality and possibly degrade the precordials. Because the true
       unipolar device records limb components, noise affecting one of the
       limbs can be evaluated beforehand, and hence operators can decide not
       to use the WCT if it is compromised without experiencing loss of the
       entire set of precordials. To this extent, the amplitude of the WCT seems
       to be dominated by the right-arm (RA) component (which is the largest
       component observable from Figure 5(b)); similar observations were
       made for the other subjects enrolled in our pilot study and hence we can
       confirm the previous hypothesis that WCT may impair chest exploration
       due to biasing imposed by the right arm 14..

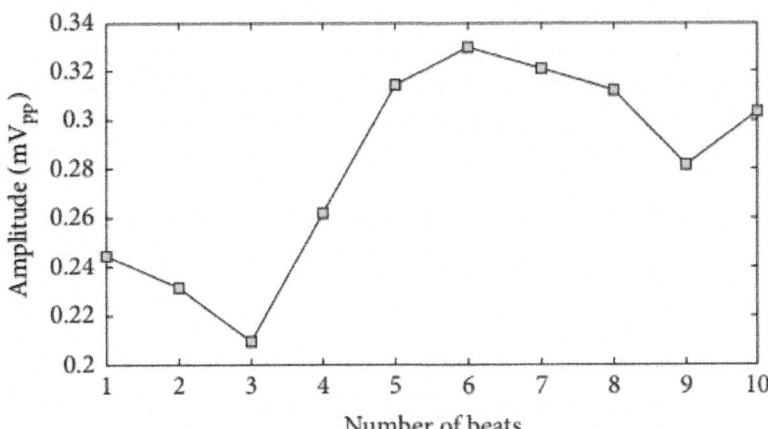

Figure 4: Variation in WCT amplitude measured across 10 consecutive beats selected
starting from a random beat within the recording (see text).

(5)   The position of the limb electrodes directly affects the shape of the
       leads and WCT. A simple comparison of Figures 5 and 6 reveals that
       the QRS feature of the WCT is distorted. When electrodes are moved to

the shoulders and hips (see Figure 6), the S-wave decreases in favor of a larger R-wave and this is particularly visible in Lead III, where the QRS is clearly larger.

(6)   In unipolar components, there is a marked increase in the amplitude of the LL component and a reversion of the LA component polarity. For these reasons it is possible to say that the increase of information carried by the lower body (LL) and the simultaneous distortion of the information carried by the upper body (LA) justify the deviation of the cardiac axis in favor of more vertical directions, as observed in literature 24. This finding is supported by an intuitive analysis of the correct formula for the calculation of the cardiac axis. Recalling that the cardiac axis is calculated by 36.

$$\text{Cardiac Axis} = \pm \tan^{-1} \frac{aV_F}{I}$$

(4)

which can be expressed in unipolar components as 28.

$$\text{Cardiac Axis} = \pm \tan^{-1} \frac{LL - ((RA + LA)/2)}{LA - RA},$$

(5)

it is easy to conclude that a marked increase in LL alone will increase the vertical component of the vector $\overline{P}$ representing the cardiac activity, shifting the value of its angle α towards a steeper value; one may note that a reversion of the LA polarity may also contribute to an increase of the numerator of the cardiac axis calculation formula, which, when limb electrodes are moved closer to the torso, is also always accompanied by a reduction of Lead I (the denominator), which may further increase the shift of α toward the vertical axis.

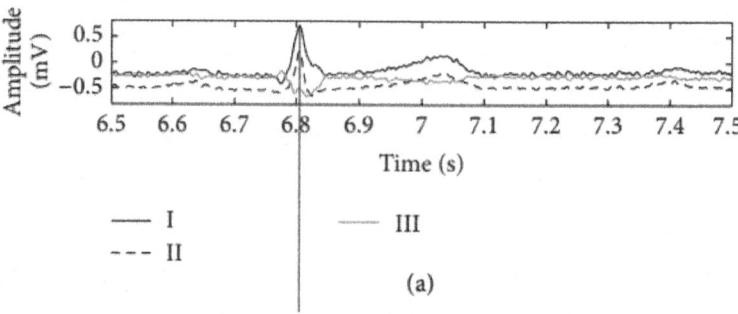

(a)

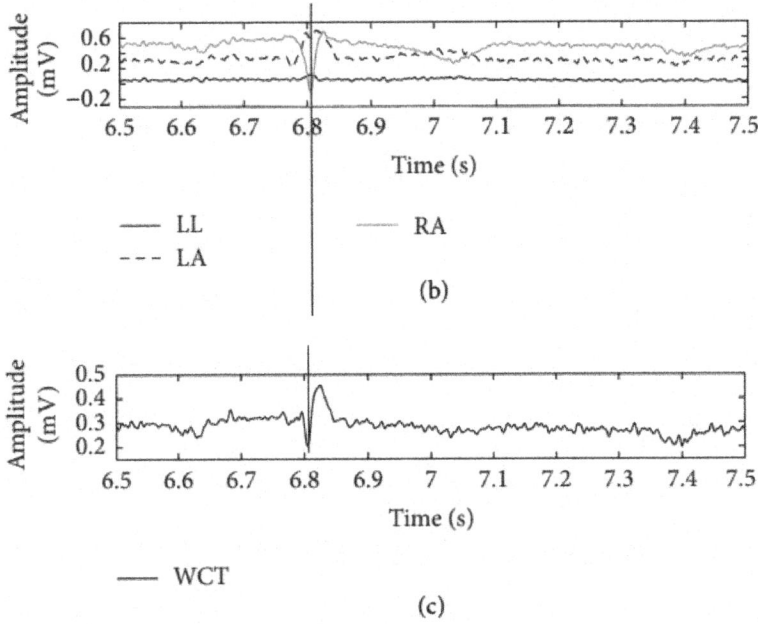

Figure 5: Direct comparison of WCT (c) with cardinal limb leads (a) and true unipolar components (b) when limb electrodes are placed on wrists and ankles. The QRS fiducial point is marked (thin vertical line) using Lead II as the reference.

Lastly, because the signals recorded with the true unipolar device are linearly independent, similar to what is done with EEG recordings, it is possible to increase the space of signals via rereferencing. Namely, the number of signal traces obtainable from the 10 placed electrodes will increase from twelve to at least thirty (nine independent unipolar ones, nine referred to the common average, and the twelve traditional signals), thereby increasing the redundancy of information present in the ECG, as has been sought since its invention more than 80 years ago 1.. In other words, a corollary of this new method is that the current practice is at the same time improved (more robustness to noise, larger redundancy of information, and visualization of WCT) and preserved (the traditional signal and diagnostic method are also useable). It is notable that reconstruction of 12-lead ECG based upon point-to-point subtraction of components can be more robust to noise. This is because signal analysts (medical practitioners annotating the ECG with or without the aid of automated procedures) will be able to estimate the signal-to-noise ratio of each individual component (such as power-line noise and artifacts) and operate individual differentiated and customized software filters on the components before reconstructing the signal 11, 12, 26, 27.

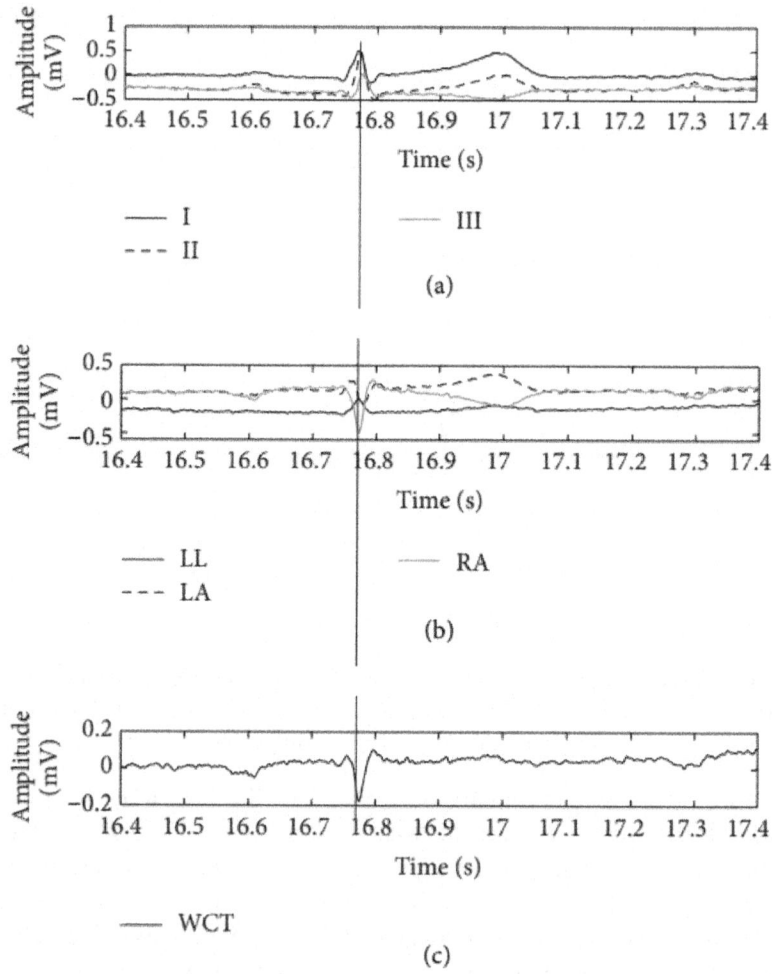

Figure 6: Direct comparison of WCT (c) with cardinal limb leads (a) and true unipolar components (b) when limb electrodes are placed on hips and shoulders. The QRS fiducial point is marked (thin vertical line) using Lead II as the reference.

CONCLUSIONS

We presented experimental evidence that the WCT is not a stable reference for ECG leads through the cardiac cycle, that its shape and amplitude (measured peak to peak) are comparable with the amplitude of other ECG leads, and most importantly that it shows clinically significant amplitude variability during the recording. With this study we also show that the WCT, like the limb leads, is directly affected by alteration of the electrode position and therefore it can pass this additional bias to precordials with unforeseen effects upon diagnosis. Using our device, in this study, we have also been able to justify the shift of the cardiac axis toward the vertical direction that has been observed in several independent studies when limb electrodes are placed closer to the torso (i.e., stress ECG). Hence, since our analysis and experiment confirm concerns about the alteration of all standard leads when limbs electrodes are placed closer to the torso, we conclude that this practice should be avoided or used only where strictly necessary (i.e., when recording is not possible otherwise).

Lastly, our technique for measurement of ECG signals, allowing calculation of the WCT and standard 12-lead ECG, offers the construction of a larger space of signals, which adds redundancy to the ECG, as has been sought since its invention more than 80 years ago 1. We are currently seeking ethical clearance for a large trial to confirm the extent and impact of our findings, particularly concerning the effect of the currently widespread practice of placing the limb electrodes closer to the torso.

REFERENCES

1.   J. Malmivuo and R. Plonsey, Bioelectromagnetism—Principles and Applications of Bioelectric and Biomagnetic Fields, Oxford University Press, 1995.

2.   R. O. Bonow, D. L. Mann, D. P. Zipes, and P. Libby, Braunwald's Heart Disease: A Textbook of Cardiovascular Medicine, Elsevier, 2012.

3.   G. D. Gargiulo, P. Bifulco, M. Cesarelli, A. Fratini, and M. Romano, "Problems in assessment of novel biopotential frontend with dry electrode: a brief review," Machines, vol. 2, no. 1, pp. 87–98, 2014.

4.   P. Horowitz and W. Hill, The Art of Electronics, Cambridge University Press, 2002.

5.   J. G. Webster, Ed., Medical Instrumentation Application and Design, John Wiley & Sons, Hoboken, NJ, USA, 2009.

6.   D. Prutchi and M. Norris, Design and Development of Medical Electronic Instrumentation, Wiley, Hoboken, NJ, USA, 2005.

7.   F. N. Wilson, F. D. Johnston, F. F. Rosenbaum, and P. S. Barker, "On Einthoven's triangle, the theory of unipolar electrocardiographic leads, and the interpretation of the precordial electrocardiogram," American Heart Journal, vol. 32, no. 3, pp. 277–310, 1946.

8.   E. Frank, "General theory of heart-vector projection," Circulation Research, vol. 2, no. 3, pp. 258–270, 1954.

9. J. E. Madias, "On recording the unipolar ECG limb leads via the Wilson's vs the Goldberger's terminals: aVR, aVL, and aVF revisited," Indian Pacing Electrophysiology Journal, vol. 8, no. 4, pp. 292–297, 2008.

10. L. Bacharova, R. H. Selvester, H. Engblom, and G. S. Wagner, "Where is the central terminal located? In search of understanding the use of the Wilson central terminal for production of 9 of the standard 12 electrocardiogram leads," Journal of Electrocardiology, vol. 38, no. 2, pp. 119–127, 2005.

11. G. Gargiulo, A. Thiagalingam, A. Mcewan, M. Cesarelli, P. Bifulco, and J. Tapson, "True unipolar ECG leads recording (without the use of WCT)," in Proceedings of the 61st Annual Scientific Meeting of the Cardiac Society of Australia and New Zealand (CSANZ '13), p. S102, Gold Coast, Australia, 2013.

12. G. D. Gargiulo, A. L. McEwan, P. Bifulco et al., "Towards true unipolar ECG recording without the Wilson central terminal (preliminary results)," Physiological Measurement, vol. 34, no. 9, pp. 991–1012, 2013.

13. E. Frank and C. F. Kay, "The construction of mean spatial vectors from null contours," Circulation, vol. 9, no. 4, pp. 555–562, 1954.

14. G. E. Dower, J. A. Osborne, and A. D. Moore, "Measurement of the error in Wilson's central terminal: an accurate definition of unipolar leads," British Heart Journal, vol. 21, pp. 352–360, 1959.

15. R. H. Bayley, E. W. Reynolds Jr., C. L. Kinard, and J. F. Head, "The zero of potential of the electric field produced by the heart beat: the problem with reference to homogeneous volume conductors," Circulation Research, vol. 2, no. 1, pp. 4–13, 1954.

16. R. H. Bayley and C. L. Kinard, "The zero of potential of the electrical field produced by the heart beat; the problem with reference to the living human subject," Circulation Research, vol. 2, no. 2, pp. 104–111, 1954.

17. N. Miyamoto, Y. Shimizu, G. Nishiyama, S. Mashima, and Y. Okamoto, "The absolute voltage and the lead vector of Wilson's central terminal," Japanese Heart Journal, vol. 37, no. 2, pp. 203–214, 1996.

18. R. H. Bayley and A. E. Schmidt, "The problem of adjusting the Wilson central terminal to a zero of potential in the living human subject," Circulation Research, vol. 3, no. 1, pp. 94–102, 1955.

19. H. C. Burger and J. B. van Milaan, "Heart-vector and leads," British Heart Journal, vol. 8, no. 3, pp. 157–161, 1946.

20. P. M. Rautaharju, R. J. Prineas, R. S. Crow, D. Seale, and C. Furberg, "The effect of modified limb electrode positions on electrocardiographic

wave amplitudes," Journal of Electrocardiology, vol. 13, no. 2, pp. 109–113, 1980.

21. O. Pahlm, W. K. Haisty Jr., L. Edenbrandt et al., "Evaluation of changes in standard electrocardiographic QRS waveforms recorded from activity-compatible proximal limb lead positions," The American Journal of Cardiology, vol. 69, no. 3, pp. 253–257, 1992.

22. D. C. Sevilla, M. L. Dohrmann, C. A. Somelofski, R. P. Wawrzynski, N. B. Wagner, and G. S. Wagner, "Invalidation of the resting electrocardiogram obtained via exercise electrode sites as a standard 12-lead recording," The American Journal of Cardiology, vol. 63, no. 1, pp. 35–39, 1989.

23. J. P. Sheppard, T. A. Barker, A. M. Ranasinghe, T. H. CluttonBrock, M. P. Frenneaux, and M. J. Parkes, "Does modifying electrode placement of the 12 lead ECG matter in healthy subjects?" International Journal of Cardiology, vol. 152, no. 2, pp. 184–191, 2011.

24. R. M. Farrell, A. Syed, A. Syed, and D. D. Gutterman, "Effects of limb electrode placement on the 12- and 16-lead electrocardiogram," Journal of Electrocardiology, vol. 41, no. 6, pp. 536–545, 2008.

25. P. Kligfield, L. S. Gettes, J. J. Bailey et al., "Recommendations for the standardization and interpretation of the electrocardiogram: part i: the electrocardiogram and its technology A scientific statement from the American Heart Association Electrocardiography and Arrhythmias Committee, Council on Clinical Cardiology; the American College of Cardiology Foundation; and the Heart Rhythm Society Endorsed by the International Society for Computerized Electrocardiology," Journal of the American College of Cardiology, vol. 49, no. 10, pp. 1109–1127, 2007.

26. G. D. Gargiulo, J. Tapson, A. Van Schaik, A. McEwan, and A. Thiagalingam, "Unipolar ECG circuits: towards more precise cardiac event identification," in Proceedings of the IEEE International Symposium on Circuits and Systems (ISCAS '13), pp. 662– 665, Beijing, China, May 2013.

27. G. D. Gargiulo, A. L. McEwan, P. Bifulco et al., "Towards true unipolar bio-potential recording: a preliminary result for ECG," Physiological Measurement, vol. 34, no. 1, 2013.

28. G. D. Gargiulo, P. Bifulco, M. Cesarelli et al., "Mean (QRS) cardiac electrical axis: a new calculation formula based on real unipolar ECG leads," in Proceedings of the Australian Biomedical Engineering Conference (ABEC '13), Sydney, Australia, 2013.

29. G. Gargiulo, R. Calvo, C. Jin et al., "Giga-ohm high-impedance FET input amplifiers for dry electrode biosensor circuits and systems," in Integrated Microsystems Electronics: Photonics, and Biotechnology, K. Iniewski, Ed., pp. 165–194, CRC Press, 2011.

30. G. Gargiulo, R. A. Calvo, P. Bifulco et al., "A new EEG recording system for passive dry electrodes," Clinical Neurophysiology, vol. 121, no. 5, pp. 686–693, 2010.

31. G. Gargiulo, P. Bifulco, A. McEwan et al., "Dry electrode biopotential recordings," in Proceedings of the Annual International Conference of the IEEE Engineering in Medicine and Biology Society (EMBC '10), pp. 6493–6496, Buenos Aires, Argentina, August-September 2010.

32. B. B. Winter and J. G. Webster, "Reduction of interference due to common mode voltage in biopotential amplifiers," IEEE Transactions on Biomedical Engineering, vol. 30, no. 1, pp. 58–62, 1983.

33. J. D. Enderle, Bioinstrumentation, vol. 6, Morgan & Claypool, 2006.

34. W. Byes, Ed., Instrumentation Reference Book, CRC Press, Boca Raton, Fla, USA, 2002.

35. H. C. Burger, "The zero of potential: a persistent error," American Heart Journal, vol. 49, no. 4, pp. 581–586, 1955.

36. D. Novosel, G. Noll, and T. F. Luscher, "Corrected formula for " the calculation of the electrical heart axis," Croatian Medical Journal, vol. 40, no. 1, pp. 77–79, 1999.

Chapter 11

IMPROVEMENT OF EEG SIGNAL ACQUISITION: AN ELECTRICAL ASPECT FOR STATE OF THE ART OF FRONT END

Ali Bulent Usakli

Department of Technical Sciences, The NCO Academy, 10100 Balikesir, Turkey

ABSTRACT

The aim of this study is to present some practical state-of-the-art considerations in acquiring satisfactory signals for electroencephalographic signal acquisition. These considerations are important for users and system designers. Especially choosing correct electrode and design strategy of the initial electronic circuitry front end plays an important role in improving the system's measurement performance. Considering the pitfalls in the design of biopotential measurement system and recording session conditions creates better accuracy. In electroencephalogram (EEG) recording electrodes, system electronics including filtering, amplifying, signal conversion, data storing, and environmental conditions affect the recording performance. In this paper, EEG electrode principles and main points of electronic noise reduction methods in EEG signal acquisition front end are discussed, and some suggestions for improving signal acquisition are presented.

INTRODUCTION

Although basics of the electroencephalogram (EEG) measurement in man have been the same since 1929, it was first made by Hans Berger, the technological developments give the opportunity to build much more sophisticated acquisition systems regarding clinical needs and scientific researches. The human brain generates electrical signals called EEG signals which are related to body functions, and this paper is about their acquiring. These signals are roughly less than $100\mu V$ and $100\,Hz$ and can be measured with electrodes placed on the scalp, noninvasively. Because of their low amplitude due to the skull's composition, the measurement of EEG is more difficult than the other noninvasive biosignal measurements such as the electrocardiogram, electromyogram, electrooculogram, and so forth. Having expensive bio-signal recording systems cannot guarantee acquiring proper signals. In that sense, some factors to acquire good EEG signals should be considered in new designs and during recording sessions. These major considerations are discussed and some suggestions are presented in this paper.

In bio-signal recordings, electrodes are the initial elements which are used for converting biopotential signals due to biopotential sources into electrical signals. Figure 1 shows the simplified biopotential measurement. EEG electrodes are usually made of metal and are produced as cup-shaped, disc, needle, or microelectrode to measure intra-cortex potentials. Silver chloride (AgCl) is preferred for common neurophysiologic applications [1]. Because Ag is a slightly soluble salt, AgCl quickly saturates and comes to equilibrium. Therefore, Ag is a good metal for metallic skin-surface electrodes [2]. Choosing the correct electrode as well as preparation of the skin before recording affects the accuracy of the measurements.

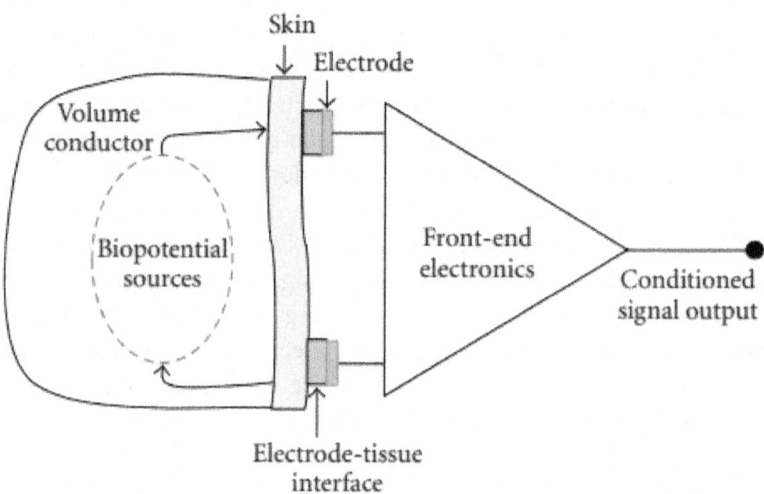

Figure 1: Biopotential measurement via electrodes.

Another major factor is electronic noise which is quite important for the bio-signal measurements. Electronic noise can be caused by internal and external noise sources. The internal noise sources are thermal (due to resistive components), shot (due to semiconductor holes and diffusions), flicker (due to contact pins), and burst (or popcorn, due to impurities in semiconductors) noise [3]. The most important external noise is caused by power-line interference. It is clearly seen in spectral analysis at 50 Hz (or 60 Hz). Between power lines and the subject there are capacitances (parasitic and isolation). To extract biosignals precisely from electronic noise requires efficient noise reduction methods [3, 4]. Efficient analog and/or digital filtering are needed for this purpose.

In the following sections, EEG electrodes as well as EEG recording and design considerations are presented.

MATERIALS AND METHODS

EEG Electrodes

Electrodes may be polarized (nonreversible) or nonpolarized (reversible). Polarization is avoided since the chloride ion is common to both the electrode and the electrolyte. Other metals such as gold or platinum can be used for electrode fabrication but is costly. Polarized electrodes tend to make significant capacitance, and this may interfere with the transmission of underlying bio-signals. These electrodes behave like a low-frequency filter (low-pass filter). Non-polarized electrodes, such as those of AgCl, are preferred for common neurophysiologic applications [1, 2]. Normal Ag/AgCl electrodes need to be chlorinated in time; however, sintered (making electrodes from powder, by heating the material in a sintering furnace below its melting point) Ag/AgCl electrodes do not need to be chlorinated.

The EEG electrodes can be classified as disposable reusable disc and cup shaped (EEG caps), subdermal needles (single-use needles that are placed under the skin), and implanted electrodes (to precisely pinpoint the origin of seizure activity). Needles are available with permanently attached wire leads, where the whole assembly is discarded, or sockets that are attached to lead wires with matching plugs. They are made of stainless steel or platinum. Some of EEG electrodes can be used for special applications. For example, implanted EEG electrodes also can be used to stimulate the brain and map cortical and subcortical neurological functions, such as motor or language function, in preparation for epilepsy surgery. Infection must be considered a major risk of implanted EEG electrodes.

In a noninvasive electrical brain signal measurement, there is an interface material between the electrode and the skin. This material is an electrolyte and can be in EEG gel or paste form. The electrophysiological activity caused by a biopotential source is a current source that causes current flow in the extracellular fluid and other conductive paths through the tissue.

A cup-shaped electrode provides enough volume to contain an electrolyte, including chlorine ions. In these electrodes, the skin never touches the electrode material directly. The electrode-tissue interface has impedance depending on several factors. Some of these factors are the interface layer (such as skin preparation, fat mass, hair, etc.), area of electrode's surface, and temperature of the electrolyte. The electrode-tissue contact can be modeled as in Figure 2. As it is seen from the figure, the electrode-tissue interface not only is resistive but consists of capacitive elements too. This is important for the frequency dependency of the electrode-skin contact.

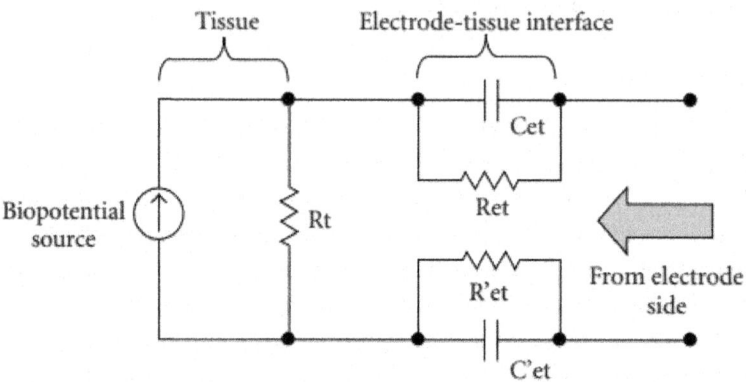

Figure 2: Simplified equivalent circuit of biopotential source and electrode-tissue interface from electrode. Biopotential source as a current source and tissue resistance is shown Rt. Cet and Ret electrode-tissue equivalent elements may change for each electrode contact.

Because of the interaction between metallic electrode and electrolyte, the ions accumulated as parallel plates. Ion-electron exchange occurs between the electrode and the electrolyte. This exchange results in voltage and it is called the half-cell potential. Because of this potential, in some cases, biopotential amplifiers must tolerate up to ± 300 mV. This value depends on the electrode and electrolyte materials. This can be explained by the Nernst Equation, simply, as

$$\varepsilon = \varepsilon^0 - \frac{0,05916}{n_e} \log Q.$$

(1)

Here, we have ε: Half-cell potential (V), n_e: Transported electron (mol number), and Q: Rate of inside and outside ions: $Q = [\text{Ions}_{\text{Inside}}]/[\text{Ions}_{\text{Outside}}]$.

In clinical EEG recordings, 10/20 Electrode Placement System is a standard and in general, it has been used for many clinical or research applications [5]. Although there are 75 locations in this system, 8 to 32 electrodes may be sufficient for clinical applications. 8 channels can also be sufficient for some

Brain Computer Interface (BCI) applications; on the other hand, for Electrical Source Imaging (ESI) more than 100 channels are required. Electrodes are positioned over the frontal, temporal, parietal, and occipital lobes, and odd and even numbers refer to the left and right hemispheres, respectively. Because of the requirements, another placement system is 5% electrode placement and 345 locations are determined [6], but it is not a common standard.

EEG Recordings

In the EEG system, as a non-invasive application, the electrodes are placed on the scalp, and a sufficient number of electrodes may be 1 to 256 (or more in near future) placed on EEG cap for easy application. To provide ionic current and to reduce contact impedance between the electrode surface and the scalp, EEG gel or paste must be used together for proper skin preparation. In biopotential measurements, the most important point is preserving the biosignal's originality. The contact impedance should be between $1\,k\,\Omega$ to $10\,k\,\Omega$ to record an accurate signal. Less than $1\,k\,\Omega$ contact impedance indicates a possible shortcut between electrodes; on the other hand, impedance greater than $10\,k\,\Omega$ can cause distorting artifacts.

Drying the gel or paste in time, as well as the subject's perspiration and movements (eye blinks, muscle movements, heart beats, etc.), can easily affect the measurement performance negatively. Because of these reasons, recording time is generally limited for several hours. For long periods of time or ambulatory EEG recordings, additional requirements are necessary to make patients more comfortable and to allow for consistent system performance. High-resolution applications such as ESI or wireless data transfer also require a different approach for the design of the novel electrodes. To reduce the skin preparation time and to measure the bio-signals more accurately, several approaches are attempted for electrode fabrication. For example, multiarray thin film electrodes are developed especially for different depth in operational applications [10]; nitride-covered steel is used as an electrode and there is no need for EEG paste to result in successful recordings [11]. In the last few years, active electrode (small or unity gain amplifier close to electrode) research is gaining popularity. With these types of electrodes, without the use of electrode gel and with much more skin preparation, noise reductions are reported [9, 12–14].

In commercial applications, apart from classical cup- or disc-shaped electrodes and active electrodes, another approach is used to reduce preparation time (by EGI's HydroCel Geodesic Sensor Net). In this approach, scalp preparation and abrasion are not required. Because the soft pedestal design of the chamber creates a sealed environment, it hydrates the skin and creates an interface between the skin and electrode. Application times for the sponge-based HydroCel Geodesic Sensor Net that range between 5 minutes for 32 channels to 15 minutes for 256 channels are reported [8]. In practical consideration, at least 45 minutes are required for the electrode while 15 minutes are reasonable in skin preparations for the 256-channel cup-shaped electrode cap. Figure 3shows some EEG electrodes and caps commercially available. In this figure several examples as non-invasive electrodes and EEG caps as well as one intracortical electrode array are shown.

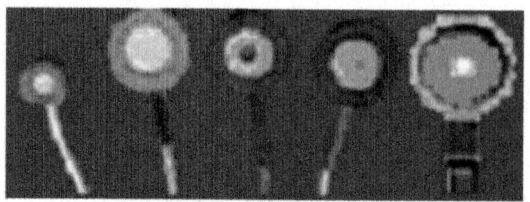

(a) Ag-AgCl electrodes

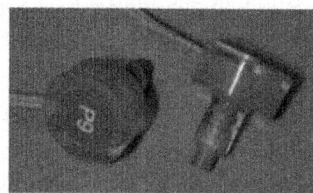

(b) Active electrodes

(c) Intra-cortical electrode array

(d) EEG caps (left to right: Standard: 256-ch., (mc) (Neuroscan) [7], (Neuroscan) [7], (EGI) [8], Active; (Biosemi) [9], Hydrocel (EGI)) [8]

Figure 3: Commercially available EEG electrodes and cap samples; (c) is for invasive applications.

Another approach for fabricating EEG electrode is dry electrode (Figure 4). This type of electrode does not need an extensive set-up time, and it is convenient for long-term recordings. These properties are advantageous for BCI and neurofeedback applications. As an example, in order to design dry electrode, 1.5 mm thick silicone conductive rubber-shaped discs of 8 mm diameter are used. The active side of the electrode is capacitive and coupled through a layer of insulating silicon rubber with a metal shield wired to the active guard shield. The impedance of the realized electrodes at 100 Hz is greater than 20 M Ω with a parasitic capacitance smaller than 2 pF [15].

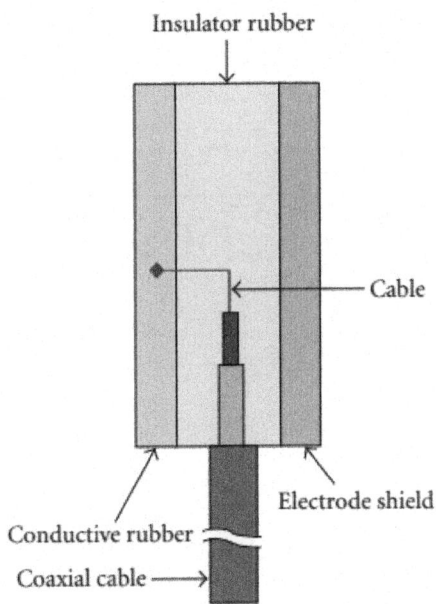

Insulator rubber

Cable

Electrode shield

Conductive rubber

Coaxial cable

Figure 4: A dry electrode principle.

For under cortex applications intra-cortical electrodes are used. One of these types of electrodes (The Utah Intra-cortical Electrode Array) is an array of 100 penetrating silicon microelectrodes designed to electrically focus stimulation or record neurons residing in a single layer up to 1.5 mm beneath the surface of the cerebral cortex [16]. Each electrode of the intra-cortical array electrode is 1.5 mm long, 0.08 mm wide at the base, and 0.001 mm at the tip.

Each type of electrode should be used with a successful electronic circuit. In Figure 5 EEG electrode and initial signal acquisition examples are shown. Recording environment conditions, contact impedance value and its stability, amplification method (ac or dc), and recording time must be considered. In the next subsection design considerations are explained, briefly.

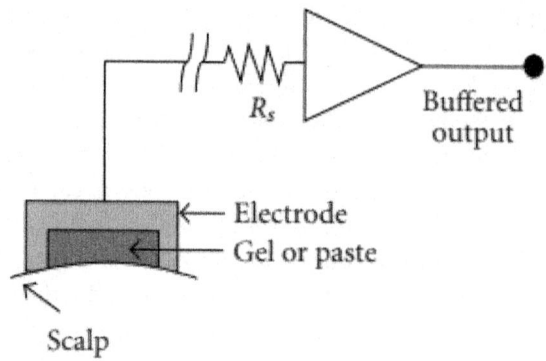

(a) A cup-shaped electrode

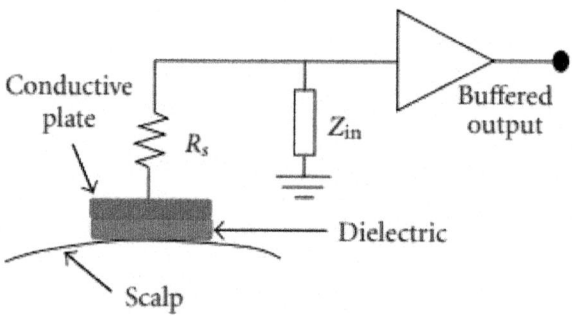

(b) Principle of capacitive electrode

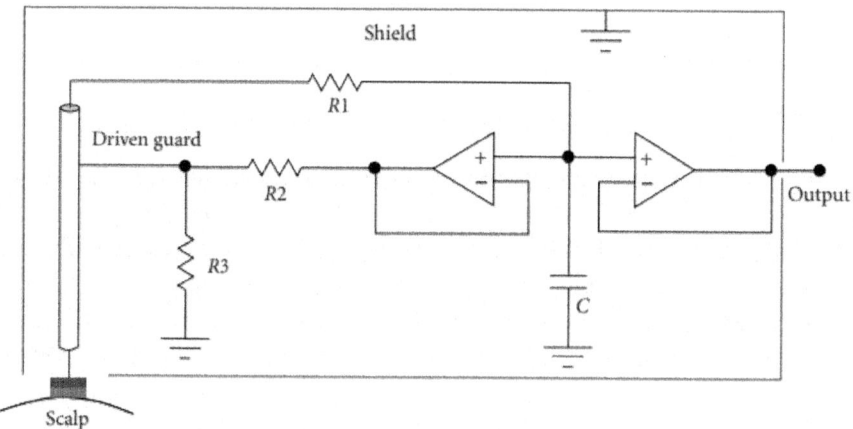

(c) Driven guard and shielding with metal plates to reduce electromagnetic interferences

Figure 5: EEG electrode connections.

The Design Considerations

In EEG recordings, as the other bio-signal measurements, one of the major problems is the 50 (or 60) Hz noise due to power lines. Between power lines and the subject there are capacitances (parasitic and isolation). Electromagnetic interference (EMI) ways are shown in Figure 6. The environmental factors influence the subject and measurement system. For example, a fluorescent lamp 1-2 m away from the measurement system interferes with the measuring signal as several kHz peaks. The interference signal may be half of the power line noise. In the same way, other electrical or electronic devices may interfere with the bio-signal measurements.

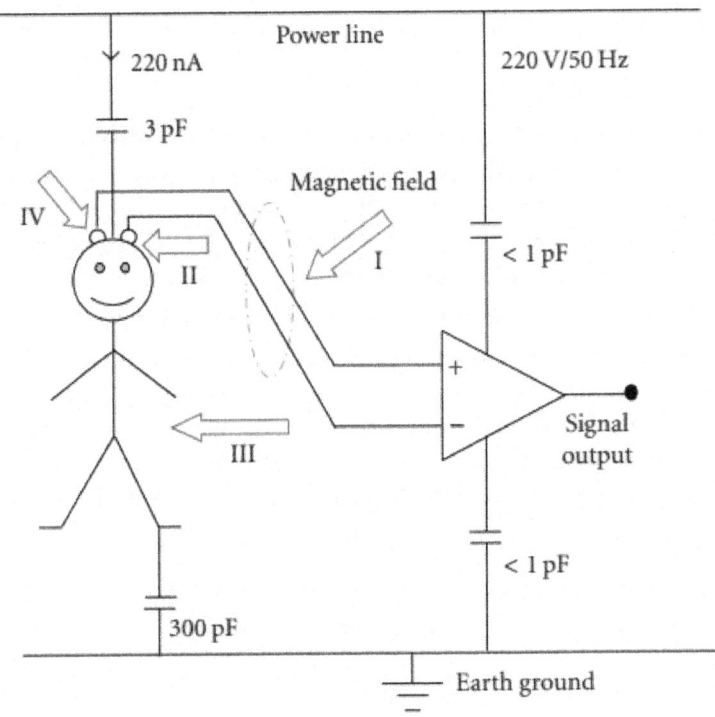

Figure 6: Electromagnetic interference (EMI) ways (for the capacitance values [9]). Arrows show the interference currents. (I) Voltage due to magnetic field to electrode cable loop is illustrated. (II) Displacement current on subject head due to electrical field causes voltage drop across electrodes. (III) Displacement current on subject body due to electrical field causes voltage drop across electrodes. (IV) Additionally, this current causes voltage between measurement electrode and amplifier common pin [24].

The dc component of the common mode signal is about several thousand, volts and the ac component may be about 1 V. This value may be in mV scale when the subject's body is grounded with earth ground and may be as high as 20 V when power line is held [17]. Electrostatic discharge (ESD) and defibrillator should be considered in electronic design. Protection for these must be provided for patient/subject and also initial active components. To reduce common mode signal effects, the instrumentation amplifiers having higher common mode rejection ratio (CMRR) must be used [18, 19]. Some research studies related to biopotential design report that 80–136 dB CMRRs are obtained [20–23]. To reduce isolation capacitance effects, battery powered operation is efficient [9].

In order to guarantee the subject/patient safety, leakage current should be less than the levels determined by IEC 601-1. According to this regulation, leakage current must be less than 10 μA in normal conditions, while regarding the connection to the main power supply, 50 μA is allowable.

Biopotential amplifiers can be dc coupled or ac coupled. In design strategy, dc or ac coupling and filtering (hardware or software) decisions are the initial steps for the biopotential measurement system designer. For ac amplification more than 10 bits digital resolution may be enough. However, for dc amplification, because number of effective bits is decreased, more than 20 bits are necessary for the analog digital converter (ADC). For high digital resolution sigma-delta technology ADC is one of good solutions [22, 23].

The input amplifier circuit is presented in Figure 7. It can be used wired electrodes (classical approach) or close to electrodes (active electrode approach). After the tests, it is observed that the circuit can be used for EEG, EOG or EMG signal measurements. This amplifier's gain is 16, and it does not cause biopotential amplifier saturation, and its CMRR is 102 dB under no shielding conditions or EMI protection.

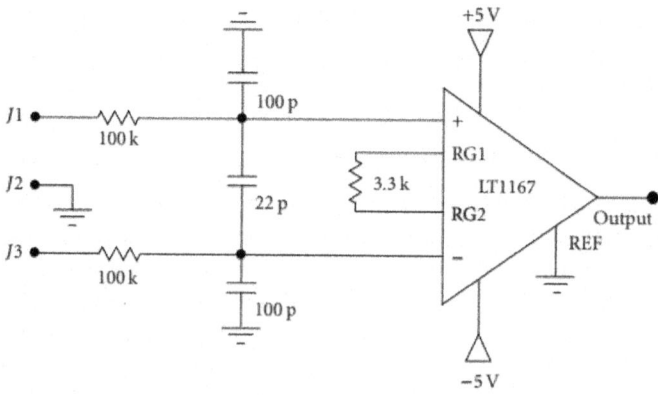

Figure 7: Suggested input amplifier circuit.

DISCUSSIONS

Acquiring EEG signal properly means mainly safety, bio-signal measurement with higher Signal to Noise Ratio (SNR) and no data loss. The major points that the author briefly proposed for the entire recording process are the following.

(i)   Subject/Patient Safety. Because of the leakage current from system electronics and defibrillator (if used), subject/patient safety should be provided. Subject/patient and front-end circuitry and earth grounds should be separated (i.e., analog and digital grounds). Increasing the isolation mode rejection ratio of the amplifier reduces the influence of isolation mode voltage.

(ii)   EMI Protection. Operation of electrical or electronic devices and especially fluorescent lamps near the recording set-up is prohibited. Otherwise acquired EEG signals are distorted and the signal corrupted with noise. Using instrumentation amplifier can help getting rid of this problem.

(iii)   No Subject/Patient Muscular Movements. Muscular movements (i.e., EMG-related contamination) such as eye blinks, clenching teeth, movement of shoulders or legs, and so forth affect EEG signal acquisition, badly. These cases may cause wrong comments on the signals and signal processing error.

(iv)   ESD Protection. Active electronic components must have greater than 2000 V ESD protection. No ESD protection may cause damage of active electronic components and may cause serious problem for subject/patient.

(v)   Efficient Grounding. Metal cases must be connected to metal plate/rod buried under ground. Proper grounding technique helps to reduce noise therefore increasing SNR.

(vi)   Electrodes. Choosing correct electrode and montage should be decided regarding clinical or research application purposes. In addition to availability of commercial standard or active types, electrodes can be made such as capacitive coupling or dry electrode. Number of electrodes and their placement is also important for the application.

(vii)   Electrode Contact Impedance. Contact impedance value must be between $1\,k\,\Omega$ and $10\,k\,\Omega$ for classical electrodes. Less than $1\,k\,\Omega$ contact impedance indicates probable shortcut between the electrodes. Greater than $10\,k\,\Omega$ contact impedance prevents acquiring EEG signals. Before measurement, contact impedance should be measured, and EEG trace should be observed while recording. Using todays technology, high input impedance ($>1\,G\,\Omega$) amplifier chips and active electrode approaches decrease dependency of the contact impedance. To acquire proper signal, electrodes should not be moved. Otherwise it causes fluctuation of the EEG signal, and spikes on it.

(viii) Noise Immunity. Noise reduction techniques must be considered in electronic circuitry and printed circuit board design. Electronic cards and connection cables should be placed in a metal box to reduce electronic noise as much as possible. Using twisted, blended, and driven signal cables gives good results. Because EEG signals are low amplitude μVs, they are very sensitive to electronic noise. Electronic noise should be less than 2 μV (peak-to-peak).

(ix) Environmental Conditions. If the EEG system is combined with magnetic resonance imaging, it must be compatible for operating under a high magnetic field. Similarly, if the EEG system is used in an operation room while surgery, it should not be affected under high electronic noise conditions such as electro-cautery. Recording must be stable and ambient temperature should not affect system performance. System performance should be independent of reasonable temperature fluctuations.

(x) Reduction for Common Mode Signals. To reduce common mode signals, it is necessary to use instrumentation amplifier having greater than 80 dB CMRR. This is important for high SNR signal acquisition.

(xi) Recording Mode. Designer or user should decide the recording mode regarding their applications. EEG recordings can be bipolar (differentiated two close electrode signal), unipolar (differentiated common reference), or using averaged reference. Clinical or research applications require different recording modes.

(xii) Reference and Ground Electrode Position. Especially reference electrode placement is important for some applications. In general reference electrode is placed on vertex; ground electrode is placed on left or right (or together) ear.

(xiii) Preserving the Biosignal Originality. Linear and distortionless amplification must be provided. Otherwise signal processing (detection, pattern recognition, feature extraction, classification, and averaging, etc.) performances may decrease.

(xiv) Avoiding Amplifier Saturation. If the amplifier circuits saturate, analog signal loss is inevitable. Amplifier saturation is caused by high input signal, mainly due to electrode movements. In ac amplification, amplifier which is used before high-pass filtering must be fixed for optimum gain to avoid saturation level. If dc amplification is preferred, there is no amplifier saturation risk; on the other hand, number of efficient bit resolution is decreased.

(xv) Cross-Talk Rejection. For multichannel systems, cross-talk rejection must be high enough. Otherwise, channels can be affected from other channels; therefore, artifact exists.

(xvi) Input Impedance. Input impedance of the circuit must be high enough. For dc signals greater than 1 G Ω input impedance value gives good results. Low input impedance causes load of bio-signal source, and it causes damaging of the signal.

(xvii) Input Bias Current. Input bias current of the input amplifier must be as low as possible (pA). If bio-signal sources are loaded by input amplifier, it also causes distortion.

(xviii) Frequency Band. Selecting the proper filter band (at least band width must be 0.5 Hz–70 Hz) is important to acquire signal. This is also important for digitizing and data storing. Sufficient (and optimum) sampling rate (>140 Hz) and transfer rate should be provided. Dc level should be removed for efficient signal conditioning/processing in hardware or in software.

(xix) Digitization. Sufficient (and optimum) digital resolution (>10 bits for ac amplification, >20 bits for dc amplification) must be provided for analog to digital converter (ADC). If low digital resolution is used, the quantization error increases.

(xx) Same Sampling Instants. If the multichannel system is designed (or used), there must be no time delay between channels. For an analog multiplier, this may be a problem, however, not for a digital multiplier. To sample at the same time instants, sample and hold circuits should be used. If each analog channel has its own ADC, this can be made with ADC control signal timing.

(xxi) Recording Time. Sufficient (and optimum) recording time is necessary. Long-time recording (more than 2 hours) causes artifacts due to drying gel, perspiration and creating of anxiety in subject/patient. On the other hand, insufficient recording time causes insufficient data acquisition.

(xxii) User Friendly. The system hardware and software must be well integrated and must have user friendly interface.

(xxiii) Low Power Consumption. This is important for especially battery-powered systems.

(xxiv) Low Cost. The system should be cost effective, and components must be available.

CONCLUSION

There are major points that should be considered to improve the measurement performance in the design of the bio-signal measurement system or recording session. Specifically choosing the correct electrode, skin preparation, and reduction of power line noise are the important issues for EEG recordings. To reduce electromagnetic interferences, a metal box for electronic circuits, a shielded (Faraday cage principle) recording room, and guarding (driven or not) for common mode signal reduction are the efficient methods. The performance of the bio-signal measurement system depends on the electrodes, electronic circuitry, and recording conditions. Choosing the correct electrode and successful electronic design strategy are essential to acquire EEG signals, properly.

REFERENCES

1. C. M. Sinclair, M. C. Gasper, and A. S. Blum, "Basic electronics in clinical neurophysiology," in The Clinical Neurophysiology Primer, A. S. Blum and S. B. Rutkove, Eds., pp. 3–18, Humana Press, Totowa, NJ, USA, 2007.

2. D. Prutchi and M. Norris, Design and Develeopment of Medical Electronic Instrumentation, John Wiley & Sons, Hoboken, NJ, USA, 2005.

3. M. Leach Jr., "Fundamentals of low-noise analog circuit design," Proceedings of the IEEE, vol. 82, no. 10, pp. 1515–1538, 1994.

4. H. W. Ott, Noise Reduction Techniques in Electronic Systems, John Wiley & Sons, New York, NY, USA, 2nd edition, 1988.

5. H. H. Jasper, "The ten–twenty electrode system of the international federation," Electroencephalography and Clinical Neurophysiology, vol. 10, pp. 367–380, 1958.

6. R. Oostenveld and P. Praamstra, "The five percent electrode system for high-resolution EEG and ERP measurements," Clinical Neurophysiology, vol. 112, no. 4, pp. 713–719, 2001.

7. Biosemi, Product catalog, 2009, http://www.biosemi.com/active_electrode.htm.

8. Neuroscan, Product catalog, 2009, http://www.neuroscan.com/quick_caps.cfm.

9. EGI, Product catalog, 2009, http://egi.com/research-division-research-products/sensor-nets.

10. O. J. Prohaska, F. Olcaytug, P. Pfundner, and H. Dragaun, "Thin-film multiple electrode probes: possibilities and limitations," IEEE

Transactions on Biomedical Engineering, vol. 33, no. 2, pp. 223–229, 1986.

11. B. A. Taheri, R. T. Knight, and R. L. Smith, "A dry electrode for EEG recording," Electroencephalography and Clinical Neurophysiology, vol. 90, no. 5, pp. 376–383, 1994.

12. W. J. R. Dunseath and E. F. Kelly, "Multichannel PC-based data-acquisition system for high-resolution EEG," IEEE Transactions on Biomedical Engineering, vol. 42, no. 12, pp. 1212–1217, 1995.

13. A. C. MettingVanRijn, A. P. Kuiper, T. E. Dankers, and C. A. Grimbergen, "Low-cost active electrode improves the resolution in biopotential recordings," in Proceedings of the 18th Annual International Conference of the IEEE Engineering in Medicine and Biology, pp. 101–102, Amsterdam, The Netherlands, October-November 1996.

14. G. Litscher, "A multifunctional helmet for noninvasive neuromonitoring," Journal of Neurosurgical Anesthesiology, vol. 10, no. 2, pp. 116–119, 1998.

15. G. Gargiulo, P. Bifulco, R. A. Calvo, M. Cesarelli, C. Jin, and A. van Schaik, "A mobile EEG system with dry electrodes," in Proceedings of IEEE-BIOCAS Biomedical Circuits and Systems Conference (BIOCAS '08), pp. 273–276, Baltimore, Md, USA, November 2008.

16. P. J. Rousche and R. A. Normann, "Chronic recording capability of the Utah intracortical electrode array in cat sensory cortex," Journal of Neuroscience Methods, vol. 82, no. 1, pp. 1–15, 1998.

17. B. B. Winter and J. G. Webster, "Reduction of interference due to common mode voltage in biopotential amplifiers," IEEE Transactions on Biomedical Engineering, vol. 30, no. 1, pp. 58–62, 1983.

18. R. Aston, Principles of Biomedical Instrumentation and Measurement, Prentice Hall, Columbus, Ohio, USA, 1990.

19. J. G. Webster, Medical Instrumentation, Application and Design, Houghton Mifflin, Boston, Mass, USA, 2nd edition, 1992.

20. G. Litscher, "A multifunctional helmet for noninvasive neuromonitoring," Journal of Neurosurgical Anesthesiology, vol. 10, no. 2, pp. 116–119, 1998.

21. I. Ulbert, E. Halgren, G. Heit, and G. Karmos, "Multiple microelectrode-recording system for human intracortical applications," Journal of Neuroscience Methods, vol. 106, no. 1, pp. 69–79, 2001.

22. A. B. Usakli and N. G. Gencer, "Performance tests of a novel electroencephalographic data-acquisition system," in Proceedings of

the 5th IASTED International Conference on Biomedical Engineering (BioMED '07), pp. 253–257, Innsbruck, Austria, February 2007.

23. A. B. Usakli and N. G. Gencer, "USB-based 256-channel electroencephalographic data acquisition system for electrical source imaging of the human brain," Instrumentation Science & Technology, vol. 35, no. 3, pp. 255–273, 2007.

24. A. B. Usakli and S. Gurkan, "Design of a novel efficient human-computer interface: an electrooculagram based virtual keyboard," Accepted in IEEE Transactions on Instrumentation and Measurement.

Chapter 12

APPLICATION CONCEPT OF ZERO METHOD MEASUREMENT IN MICROWAVE RADIOMETERS

Alexander V. Filatov

Tomsk State University of Control Systems and Radio Engineering, Tomsk, Russia

ABSTRACT

This article examined in detail microwave radiometer functioning algorithm with synchronously using of the two types of pulse modulation: amplitude pulse modulation and pulse-width modulation. This allows a zero-radiometer measurement method to realize when the fluctuation effect of the receiver gain and the influence of its own noise changes are minimized. A zero balance automatically maintains in radiometer. The antenna signal is indirectly determined through the signal duration that controls the pulse-width modulation. An analytical expression of the fluctuation sensitivity was obtained in a general form. From its analysis gain in sensitivity, conditions were defined by the optimizing of the radiometer input knot's construction. Three modifications of the radiometer input knot were researched. Fluctuation sensitivity at different measurement range was determined for modification of the radiometer input knot.

INTRODUCTION

It is well known that study of microwave appearance of various objects of earth's surface provides fundamentally different physical self-descriptiveness than using only optical and infrared earth remote sensing [1] [2] . This fact encourages continuous development of measurement systems aimed at improving the accuracy and the increasing of informative sensing saturation. A development of microwave remote sensing systems with low energy consumption intended for working in natural conditions and especially on the side refers to a difficult problem. Traditional approaches often lead to mutually exclusive solutions; therefore, the implementation of stringent requirements for the microwave radiometers is impossible without searching for new approaches, methods and solutions.

Methods for stabilizing or accounting of radiometer technical parameters' changes consist in application of different functioning methods. Among various schemes, modulation radiometers have been widely used [3] [4] . These radiometers are based on method of differential measurements. At the entrance, before radiometric receiver, modulation is generated with a certain frequency antenna signal and a stable reference noise generator signal—the antenna simulator connected to the input of the receiver, instead of the antenna. Because measurements are transferred to a higher frequency with which signals are modulated, the influence of two major destabilizing factors reduces: the influence of anomalous fluctuations of the gain near zero frequency significantly decreases; constant and quasi constant of self-noise components of the radiometer decrease. Wide application of modulation radiometers is associated with a satisfactory accuracy of measurements, which is achieved essentially by simple methods (modulation at the entrance and demodulation—synchronous detection—at the output) and a simple circuit implementation. Modulation radiometers are attracted by simple design; therefore, they are promising for repeating. This is manifested in their massive using. However, the full minimizations of effect of changes of amplifier gain and self-noise of receiver in modulation scheme do not happen. It is useful noticeably to minimize these changes by using the zero method of measurement in modulation radiometer. This method has the highest potential facilities to create precision radiometers.

Ryle and Troitsky are the authors of the conception and interpretation of zero method applying to radiometers [5] [6] . A number of successful researches have associated with the creation of zero-radiometer. These researches are proof of the Nyquist formula for spectral, fluctuating noise density of various materials' resistance validity, and discovery of the recombination rf spectral line emitted by highly excited atoms in radio astronomy, measuring abyssal temperature of the biological objects, etc. The output power of the noise reference generator is regulated before achieving the so-called zero balance in the measuring path. The so-called zero balance is being considered established if the same signals in different half-periods of a symmetric pulse of the measuring path are passing. Therefore, feedback following network and regulated reference noise source are presenting in zero radiometer. The first zero radiometer had analog regulation principle of zero balance. In this radiometer, in input knots precision operated microwave devices, adjustable attenuators or random-noise generators with regulated input power being applied. The demands of the high linear adjusting characteristic, large dynamic range, elevated response speed of regulation for bringing the measuring system in the zero balance method were established to current knots. Errors arising from using the given elements in input blocks were not allowed to fully realize the zero measurement method advantages and zero radiometers were not widely spreading.

The creation of the pulse added noise operation in the radiometer appeared as a successful development of zero method application. This operation works on pulse-width-law [7] - [9] . In this case, the average of the half- cycle modulation power of invariable reference noise signal generator is regulated by changing of the duration of its action. This led to the simplification of the input receiving block scheme (microwave switch and noise generator), improving of the linearity of calibration radiometer equation by the simple adjustment of the reference signal.

This method of mixing the reference and measured signals (in different ways pulse-modulated) led to a considerable simplification of input block design. But this method complicated modulated signals conditioning after square-law detector. It led to an increase in measurement error. Thus, schemes of the zero radiometer with pulse added noise turned out more difficult than schemes of the ordinary modulation radiometers, and therefore they were not widely used.

In this article, a new modification of the zero functioning principle of microwave radiometers by weak signal changing was considered, and the possibility of fluctuation sensitivity and stability of these systems were analyzed.

THE ZERO METHOD MODIFICATION OF THE SIGNAL RECEPTION

The equality of low-energy signals at the radiometric receiver input at different half-periods of a symmetric pulse modulation is achieved by the pulse duration changing. This impulse controls the introduction of additional noise signal into the supporting or the antenna paths of modulation radiometer. Time diagrams which explain the combine modulation principle in radiometer are shown in Figure 1. The control pulse-width signal t_{pws} changes in the range from zero to the half-period length t_{mod} of main modulator work.

The signals energy equality is the condition of specified balance in radiometer. These signals enter to the receivers input at different half-periods of symmetrical modulation. In Figure 1(a), these energies are proportional to the corresponding shaded areas $Q_1(t)$ and $Q_2(t)$. The signals energy equality is continuously maintained by automatic pulse-width signal duration t_{pws} changing of radiometer controlling system.

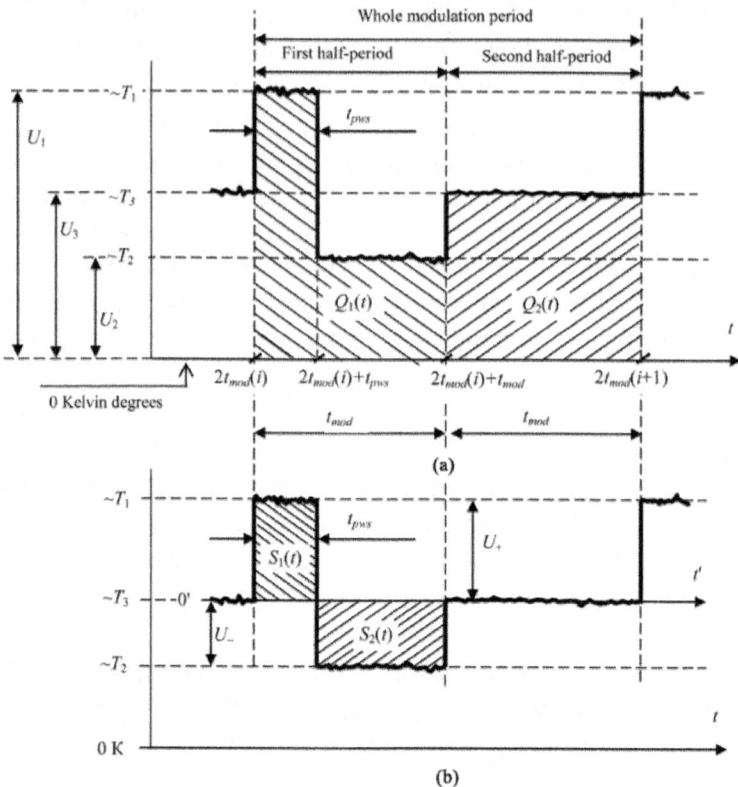

Figure 1: Signal time diagrams at the receiver output of zero radiometer which uses pulse-amplitude and pulse- width modulations.

On the diagram pulse voltage amplitudes U_1, U_2, U_3 at the receiver output are proportional to appropriate signal noise temperatures T_1, T_2, T_3 at the receiver input. Proportionality constant $G(t)$ is equal to product of gain coefficient at high and low frequencies and square-law detector transmission coefficient.

Introduce the value DQ(t) which considers the inequality of $Q_1(t)$ and $Q_2(t)$ and includes both a constant component DQ_0 and the fluctuating part of Dq(t)

$$\Delta Q(t) = Q_1(t) - Q_2(t) = \sum_{i=0}^{\infty} \int_{2t_{mod}(i)}^{2t_{mod}(i)+t_{pws}} \left[T_1 G(t-\theta) + n_1(t-\theta) \right] H_1(\theta) d\theta$$

$$+ \sum_{i=0}^{\infty} \int_{2t_{mod}(i)+t_{pws}}^{2t_{mod}(i)+t_{mod}} \left[T_2 G(t-\theta) + n_2(t-\theta) \right] H_2(\theta) d\theta$$

$$- \sum_{i=0}^{\infty} \int_{2t_{mod}(i)+t_{mod}}^{2t_{mod}(i+1)} \left[T_3 G(t-\theta) + n_3(t-\theta) \right] H_3(\theta) d\theta,$$

$$\tag{1}$$

where $G(t) = G_0 + g(t)$, G_0—constant component, $g(t)$—transmission coefficient fluctuation, $n_1(t)$, $n_2(t)$, $n_3(t)$— signal T_1, T_2, T_3 noise entries, which take into account their noise character; $H_1(q)$, $H_2(q)$, $H_3(q)$—pulse response characteristics of accumulative receiver filters for each of the corresponding signals.

For a long time interval it is possible to use $n_1(t) = n_2(t) = n_3(t) = 0$ for mean value t_{pws} considering even deviations of noise signal components which represent the normal static ergodic processes. Also for the transmission coefficient $G(t)$, taking into account the statistical invariance of $g(t)$ for a long time interval its fluctuating component can be considered equal to zero. The accumulation of signals occurs by using of three integrating RC- circuits of the first order. Pulse characteristics of these circuits are determined by the well-known relation $H(q) = \exp(-q/t)/t$, where t is circuit time constant; $t = RC$.

Then for the constant component of the parameter DQ(t), using (1) write

$$\Delta Q_0 = \sum_{i=0}^{\infty} \int_{2t_{mod}(i)}^{2t_{mod}(i)+\overline{t_{pws}}} T_1 G_0 \frac{\exp\left(-\dfrac{\theta}{\tau_1}\right)}{\tau_1} d\theta + \sum_{i=0}^{\infty} \int_{2t_{mod}(i)+\overline{t_{pws}}}^{2t_{mod}(i)+t_{mod}} T_2 G_0 \frac{\exp\left(-\dfrac{\theta}{\tau_2}\right)}{\tau_2} d\theta$$

$$-\sum_{i=0}^{\infty} \int_{2t_{mod}(i)+t_{mod}}^{2t_{mod}(i+1)} T_3 G_0 \frac{\exp\left(-\dfrac{\theta}{\tau_3}\right)}{\tau_3} d\theta.$$

$$(2)$$

It is possible to use the decomposition of exponential functions in a Maclaurin series with the two members approaching, because pulse durations t_{pws} and t_{mod} are much less than the accumulation signals time (these signals are defined by low-frequency filters constants t_1, t_2, t_3). The solution of Equation (2) is

$$\Delta Q_0 = G_0 \frac{T_1 \overline{t_{pws}} + T_2 \left(\overline{t_{mod}} - \overline{t_{pws}}\right) - T_3 t_{mod}}{2t_{mod}}.$$

$$(3)$$

Taking into account signal noise character and availability of the receiver amplifier gain fluctuation $Q_1(t)$ and $Q_2(t)$ are not equal for a modulation period. But for a large time interval in the limit for an endless number of modulation periods $DQ_0 = 0$ can be written due to automatic signals energy tracking system operation for uninterrupted leveling.

It results from (3) that

$$T_1 \overline{t_{pws}} + T_2 \left(\overline{t_{mod}} - \overline{t_{pws}}\right) - T_3 t_{mod} = 0.$$

$$(4)$$

Solving Equation (4) relatively $\overline{t_{pws}}$ obtain

$$\overline{t_{pws}} = \frac{T_3 - T_2}{T_1 - T_2} \times t_{mod}$$

(5)

Equation (5) is a mathematic of the implementation of the proposed modification of the zero method reception. It follows that the antenna signal can be determined indirectly through the added noise pulse duration without signals changing in the low frequency path. Equation (5) does not include the transfer constant of the measuring path. This indicates of zero method of radiometer working.

ZERO METHOD IMPLEMENTATION COURSE IN MICROWAVE RADIOMETERS

The comparison of $Q_1(t)$ and $Q_2(t)$ is replaced by equivalent comparison of volt-second areas (Figure 1(b)) of positive $S_1(t)$ and negative $S_2(t)$ pulses at the receiver output. The pulse amplitudes are proportional to the signal differences $T_1 - T_3$, $T_3 - T_2$ at the receiver input. If the voltage of second half-period modulation is equal to zero and time zero axis passes through the T_3 level signal, the volt-second pulse areas in the modulation first half-period are equal, $S_1(t) = S_2(t)$ (periodic sequence). As

$$S_1(t) = U_+ \overline{t_{pws}} \text{ and}$$

where $U_+ = G_0 kdf(T_1 - T_3)$ and $U_- = G_0 kdf(T_3 - T_2)$, then

$$G_0 kdf(T_1 - T_3)\overline{t_{pws}} = G_0 kdf(T_3 - T_2)\left(t_{mod} - \overline{t_{pws}}\right)$$

where k—Boltzmann constant, df—frequency range band. Solving the last equation relatively $\overline{t_{pws}}$ obtain (5).

A sequence of simple but necessary operations follows from this reasoning. These operations must be done to transform the signals after square-law detector and low-frequency amplification. These operations are constant component exclusion in signals and voltage sign determination in the second half period of modulation. The direct component exception reduces to the pulse signal T_3top deflection to the zero time axis (t Þ t'). The zero balance condition (the voltage lack in the modulation second half-period on the output of the measuring tract) settles by the t_{pws} duration regulation. Thus antenna signal tracking is realized by the t_{pws} duration changing. This leads to the signal periodic sequence displacement up or down relatively the time zero axes.

THE FLUCTUATION SENSITIVITY ANALYSIS

DQ(t) chaotic changes in (1) are associated with the parameter Dq(t) with nonzero components $n_1(t)$, $n_2(t)$, $n_3(t)$, g(t). Compute the parameter Dq(t) dispersion by correlation function method. Meanwhile take into account the statistical independence of the transmission coefficient fluctuations of the radiometer measuring path g(t) and signals T_1, T_2, T_3 noise components $n_1(t)$, $n_2(t)$, $n_3(t)$. The total dispersion is the sum of two dispersions. First of them $\overline{\Delta q_g^2}$ allows for the receiver transmission coefficient fluctuations, the second one $\overline{\Delta q_n^2}$ is caused by the noise signal nature.

The noise correlation time of function n(t) is determined by signal receiving bandwidth df and for radiometers is considerable less than modulation period $2t_{mod}$. Consequently, these signals T_1, T_2, T_3 noise components $n_1(t)$, $n_2(t)$, $n_3(t)$ are uncorrelated with each other. Then the dispersion $\overline{\Delta q_n^2}$ is determined from (1)

$$\overline{\Delta q_n^2} = \sum_{i=0}^{\infty}\sum_{j=0}^{\infty} \int_{2t_{mod}(i)}^{2t_{mod}(i)+t_{pws}} \int_{2t_{mod}(j)}^{2t_{mod}(j)+t_{pws}} \overline{n_1(t-\theta)n_1(t-\theta')} H_1(\theta)H_1(\theta')d\theta d\theta'$$
$$+ \sum_{i=0}^{\infty}\sum_{j=0}^{\infty} \int_{2t_{mod}(i)+t_{pws}}^{2t_{mod}(i)+t_{mod}} \int_{2t_{mod}(j)+t_{pws}}^{2t_{mod}(j)+t_{mod}} \overline{n_2(t-\theta)n_2(t-\theta')} H_2(\theta)H_2(\theta')d\theta d\theta'$$
$$+ \sum_{i=0}^{\infty}\sum_{j=0}^{\infty} \int_{2t_{mod}(i)+t_{mod}}^{2t_{mod}(i+1)} \int_{2t_{mod}(j)+t_{mod}}^{2t_{mod}(j+1)} \overline{n_3(t-\theta)n_3(t-\theta')} H_3(\theta)H_3(\theta')d\theta d\theta'$$
$$= J_{1n} + J_{2n} + J_{3n} \tag{6}$$

Noise n(t) autocorrelation function in comparison with pulse filters characteristics H(q) and the modulation period $2t_{mod}$ can be considered a delta function d(t) with integral value $2G_0^2 T^2$ and correlation time 1/df. Accordingly, have

$$\overline{n_1^2} = \frac{2G_0^2 T_1^2}{df}; \qquad \overline{n_2^2} = \frac{2G_0^2 T_2^2}{df}; \qquad \overline{n_3^2} = \frac{2G_0^2 T_3^2}{df}. \tag{7}$$

Exploiting (7), (6), obtain the following equation

$$\overline{\Delta q_n^2} = \frac{G_0^2}{2t_{mod}\, df} \left(\frac{T_1^2 \overline{t_{pws}}}{\tau_1} + \frac{T_2^2 \left(\overline{t_{mod} - t_{pws}}\right)}{\tau_2} + \frac{T_3^2 t_{mod}}{\tau_3} \right). \tag{8}$$

After the replacement of (5) into (8) finally obtain

$$\overline{\Delta q_n^2} = \frac{G_0^2}{2df}\left[\frac{T_1^2}{\tau_1}\times\frac{T_3-T_2}{T_1-T_2}+\frac{T_2^2}{\tau_2}\left(1-\frac{T_3-T_2}{T_1-T_2}\right)+\frac{T_3^2}{\tau_3}\right].$$

(9)

In case of equal low-frequency filters $\tau_1=\tau_2=\tau_3=\tau$

$$\overline{\Delta q_n^2} = \frac{G_0^2}{2df\tau}\left[T_3\left(T_1+T_2+T_3\right)-T_1T_2\right].$$

(10)

Next determine the dispersion caused by the fluctuations of the radiometer measuring path transmission coefficient.

Write down the formula for dispersion $\overline{\Delta q_g^2}$ calculating using the fundamental relation (1)

$$\overline{\Delta q_g^2} = \sum_{i=0}^{\infty}\sum_{j=0}^{\infty}\int_{2t_{mod}(i)}^{2t_{mod}(i)+t_{pws}}\int_{2t_{mod}(j)}^{2t_{mod}(j)+t_{pws}}T_1^2\overline{g(t-\theta)g(t-\theta')}H_1(\theta)H_1(\theta')d\theta d\theta'$$

$$+\sum_{i=0}^{\infty}\sum_{j=0}^{\infty}\int_{2t_{mod}(i)+t_{pws}}^{2t_{mod}(i)+t_{mod}}\int_{2t_{mod}(j)+t_{pws}}^{2t_{mod}(j)+t_{mod}}T_2^2\overline{g(t-\theta)g(t-\theta')}H_2(\theta)H_2(\theta')d\theta d\theta'$$

$$+\sum_{i=0}^{\infty}\sum_{j=0}^{\infty}\int_{2t_{mod}(i)+t_{mod}}^{2t_{mod}(i+1)}\int_{2t_{mod}(j)+t_{mod}}^{2t_{mod}(j+1)}T_3^2\overline{g(t-\theta)g(t-\theta')}H_3(\theta)H_3(\theta')d\theta d\theta'$$

$$+2\sum_{i=0}^{\infty}\sum_{j=0}^{\infty}\int_{2t_{mod}(i)}^{2t_{mod}(i)+t_{pws}}\int_{2t_{mod}(j)+t_{pws}}^{2t_{mod}(j)+t_{mod}}T_1T_2\overline{g(t-\theta)g(t-\theta')}H_1(\theta)H_2(\theta')d\theta d\theta'$$

$$-2\sum_{i=0}^{\infty}\sum_{j=0}^{\infty}\int_{2t_{mod}(i)}^{2t_{mod}(i)+t_{pws}}\int_{2t_{mod}(j)+t_{mod}}^{2t_{mod}(j+1)}T_1T_3\overline{g(t-\theta)g(t-\theta')}H_1(\theta)H_3(\theta')d\theta d\theta'$$

$$-2\sum_{i=0}^{\infty}\sum_{j=0}^{\infty}\int_{2t_{mod}(i)+t_{pws}}^{2t_{mod}(i)+t_{mod}}\int_{2t_{mod}(j)+t_{mod}}^{2t_{mod}(j+1)}T_2T_3\overline{g(t-\theta)g(t-\theta')}H_2(\theta)H_3(\theta')d\theta d\theta'$$

$$= J_{1g}+J_{2g}+J_{3g}+J_{4g}-J_{5g}-J_{6g}.$$

(11)

For the dispersion calculation take g(t) stationary, with a normal distribution, the exponential autocorrelation function

$$\overline{g(t-\theta)g(t-\theta')} = \sigma_g^2\exp\left(-\frac{|\theta-\theta'|}{\tau_0}\right),$$

(12)

where σ_g^2 is the measuring path transmission coefficient fluctuations dispersion, t_0—effective correlation time

constant of the transmission coefficient fluctuations. Typically t_0 is much greater than the radiometer accumulating filters time constants t_1, t_2 and t_3.

Compute generalized integral, through which any of the six integrals in expression (11) can be expressed. For variable arguments during the generalized integral calculating accept the conditions $2t_{mod} \geqslant x$, $y \geqslant 0$; $2t_{mod} \geqslant z$, $v \geqslant 0$; k, $r \gg 2t_{mod}$. Then

$$
I(x,y,z,v,k,r)
$$

$$
= \sum_{i=0}^{\infty}\sum_{j=0}^{\infty} \int_{2t_{mod}(i)+x}^{2t_{mod}(i)+y} \int_{2t_{mod}(j)+z}^{2t_{mod}(j)+v} \exp\left(-\frac{|\theta-\theta'|}{\tau_0}\right)\frac{1}{kr}\exp\left(-\frac{\theta}{k}-\frac{\theta'}{r}\right)d\theta d\theta'
$$

$$
\cong \frac{(y-x)(v-z)}{(2t_{MOD})^2 kr} \times \int_0^{\theta}\left[\int_0^{\theta}\exp\left(-\frac{\theta'}{\tau_0}\right)\exp\left(-\frac{\theta}{k}-\frac{\theta'}{r}\right)d\theta' + \int_{\theta}^{\infty}\exp\left(-\frac{\theta'-\theta}{\tau_0}\right)\exp\left(-\frac{\theta}{k}-\frac{\theta'}{r}\right)d\theta'\right]d\theta
$$

$$
= \frac{(y-x)(v-z)\tau_0}{(2t_{MOD})^2} \times \frac{2rk+k\tau_0+r\tau_0}{(r+k)(k+\tau_0)(r+\tau_0)}.
$$

(13)

Using (13) calculate the integrals $J_{1g} - J_{6g}$ in (11) and obtain an expression for calculating the fluctuation dispersions of the transmission coefficient

$$
\overline{\Delta q_g^2} = \frac{\sigma_g^2 \tau_0}{4}\left[\frac{(T_3-T_2)^2}{(T_1-T_2)^2}\left(\frac{T_1^2}{\tau_1+\tau_0}+\frac{T_2^2}{\tau_2+\tau_0}\right)+\frac{T_2^2}{\tau_2+\tau_0}\times\frac{T_1+T_2-2T_3}{T_1-T_2}+\frac{T_3^2}{\tau_3+\tau_0}\right.
$$

$$
+2T_1T_2\frac{(T_3-T_2)(T_1-T_3)}{(T_1-T_2)^2}\times\frac{2\tau_1\tau_2+\tau_1\tau_0+\tau_2\tau_0}{(\tau_1+\tau_2)(\tau_1+\tau_0)(\tau_2+\tau_0)}-2T_1T_3\frac{T_3-T_2}{T_1-T_2}
$$

$$
\left.\times\frac{2\tau_1\tau_3+\tau_1\tau_0+\tau_3\tau_0}{(\tau_1+\tau_3)(\tau_1+\tau_0)(\tau_3+\tau_0)}-2T_2T_3\frac{T_1-T_3}{T_1-T_2}\times\frac{2\tau_2\tau_3+\tau_2\tau_0+\tau_3\tau_0}{(\tau_2+\tau_3)(\tau_2+\tau_0)(\tau_3+\tau_0)}\right].
$$

(14)

If the filters pulse characteristics are the same, obtain $\overline{\Delta q_g^2} = 0$. A zero value of obtained dispersion indicates

on more less its value changing, than the dispersion caused by the signals noise nature. Given this dispersion of fluctuations of the transmission coefficient can be neglected. In fact an error of the dispersion determination can appear as results of the autocorrelation function approximate sampling (12), approximations for the exponential function. The received result is being coincided with the conclusions of other authors [10] [11] . These conclusions show that the zero method application allows minimizing the influence of fluctuations receivers increasing on the measurement results.

If pulse-width signal t_{pws} duration changing have happened at 1 time interval (discrete step), $Q_1(t)$ changing in the first modulation half-cycle (Figure 1(a)) equal to

$$
\Delta Q_{1ds} = G_0(T_1-T_2)\frac{1}{N},
$$

(15)

where N is quantity of discrete step to be placed on the half period t_{mod} duration. N characterizes measurements resolution.

The traditional sequence of operations is used to determine the sensitivity [12]. In the case of introduced null method measuring this traditional sequence of operations can be formulated as follows: the regulation of the pulse-width signal duration will be meaningful if this duration changing on one discrete step and associated with this volt-second areas changing of pulse signals at the receiver output in the first half-period modulation are equal to standard deviation from volt-second areas data equality. The standard deviation is caused by fluctuations and noise nature of the measuring signals. In accordance with this write

$$\frac{\Delta Q_{1ds}}{\sqrt{\overline{\Delta q^2}}} = \frac{1}{\sqrt{R}},$$

(16)

where R is a number of accumulated values of the pulse-width signal duration digital codes. There are two stages of averaging the signal in radiometers which use this modification of zero receiving method. Firstly a low-fre- quency analog signal filtering at the output of the receiver is performed. Further, in the system of added noise signal duration automatic control, except the adjustment cycle, the accumulation of digital codes of this duration with following averaging is occurring. It is known from the theory of errors; the signal dispersion is reduced $\left(\sqrt{R}\right)^{-1}$ times.

After the replacement of (10), (15) into (16) obtain

$$2df\tau\left(T_1 - T_2\right)^2 R = N^2\left[T_3\left(T_1 + T_2 + T_3\right) - T_1 T_2\right].$$

(17)

Pulse-width signal duration is variated from 0 to t_{mod} by changing the antenna signal from minimum to maximum. Therefore for the minimum antenna signal DT_a, which can be detected, there is a proportion

$$\frac{dT_a}{N} = \Delta T_a \sim \Delta t = \frac{t_{mod}}{N},$$

(18)

where dT_a—antenna signal measuring range, Δt—time discrete step duration, which varies by the duration t_{pws} sudden change. The value of the minimum detectable antenna signal DT_a characterizes the sensitivity.

Typically, DT_a and dT_a are defined by the device designing. In terms of these parameters, using (18) find N. N determines the number of order n of radiometer output digital code. Using (18) the formula of radiometer fluctuation sensitivity calculation is formed from (17)

$$\Delta T_a = \frac{dT_a}{\sqrt{2df\tau R}} \times \frac{\sqrt{T_3(T_1+T_2+T_3)-T_1T_2}}{T_1-T_2}.$$

(19)

For the known radiometric receiver necessary sensitivity is achieved by selecting the t and R.

Parameter R is related to measurement time t_{mes} by the ratio

$$R = \frac{t_{mes}}{2t_{mod}}.$$

STRUCTURED MODELING OF THE RADIOMETER IN-PUT RECEIVING MODULES

The signal modulation before their arrival at the receiver is carried out in the radiometer input block. Three signals are subjected to modulation. Two signals are reference signals T_{ref} and T_{add}. They are produced by reference noise generator (NG) and additional reference noise generator (ANG) accordingly. The third signal is an antenna measurable signal (A) T_a. Combinations of these signals constitute the signals T_1, T_2, T_3 on time diagrams in Figure 1. Possible signals combinations T_{ref}, T_{add}, T_a for the level formations T_1, T_2, T_3 are given in Table 1. T_n?reduced to the input of the receiver self-noise effective temperature of the receiver and input block of the radiometer.

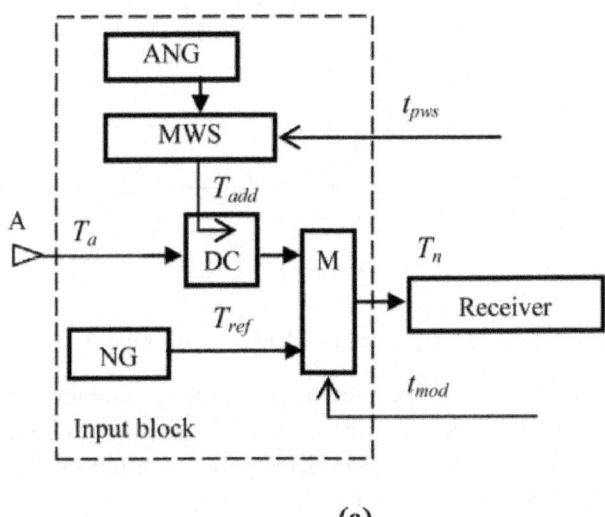

(a)

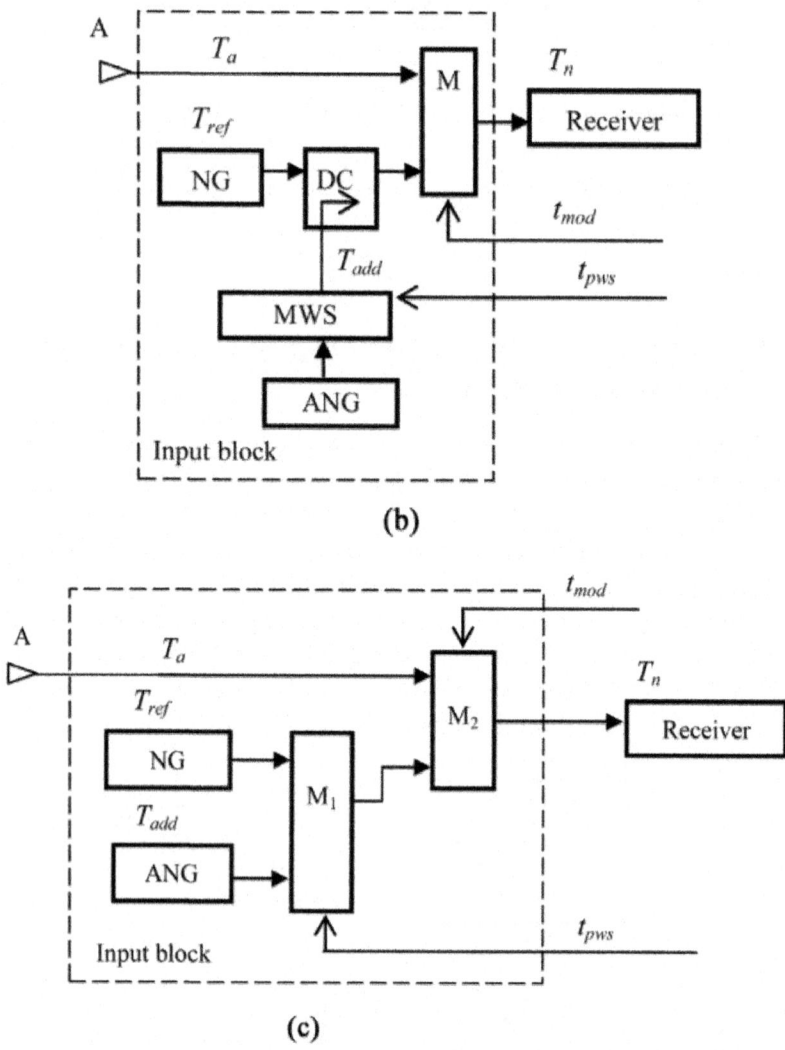

Figure 2: Input receiver block structure schemes of the modified zero radiometers.

According to the received method functioning model (5) block diagrams of the three input blocks (for the respective positions of the Table 1) are shown in Figure 2. Modulator (M) alternately commutes to the receiver input or antenna signal T_a or noise reference generator signal T_{ref}. In the schemes in Figure 2(a) and Figure 2(b) additional reference signal T_{add} is introduced through the microwave switch (MWS) to the antenna or supporting duct through the directional coupler (DC). Additional modulator (M_1) is introduced into the scheme in Figure 2(c).

Table 1: The modulated signals combinations for different measurement ranges

| Position | The modulated signals | | | Note |
	T_1	T_2	T_3	
1	$T_a + T_{add} + T_n$	$T_a + T_n$	$T_{ref} + T_a$	$T_a < T_{ref}$
2	$T_{ref} + T_{add} + T_n$	$T_{ref} + T_n$	$T_a + T_n$	$T_a > T_{ref}$
3	$T_{add} + T_n$	$T_{ref} + T_n$	$T_a + T_n$	$T_{add} > T_a > T_{ref}$

Obtain the transfer characteristic of the radiometer for the circuit of input block in Figure 2(a) by the values of effective temperatures from Pos. 1 Table 1 substituting in Equation (5)

$$t_{pws} = \frac{T_{ref} - T_a}{T_{add}} \times t_{mod}.$$

(20)

(t_{pws} duration shown without upper underlining and further averaging mark is ignored). From (20) follows that antenna signal can be determined through the pulse-width signal duration that controls the additional reference noise generator (ANG) modulation without signals change in the radiometer low-frequency part. t_{pws} duration is associated with antenna signal T_a by the linear law and does not depend on the transmission coefficient of the radiometer.

Obtain antenna signal from (20) $T_a = T_{ref} - T_{add} t_{pws}/t_{mod}$. Define the minimum and maximum of measured signals range limit by substituting two sample extremes of the duration t_{pws} (equal t_{mod} and 0) in the last equality. $T_{a,min} = T_{ref} - T_{add}$; $T_{a,max} = T_{ref}$. Where the effectives dynamic range is determined by the additional reference noise generator (ANG) signal $dT_a = T_{a,max} - T_{a,min} = T_{add}$.

The equation for sensitivity determining of radiometer with the input block Figure 2(a) can be found from (19) by substituting signals T_1, T_2, T_3 of the Pos. 1 Table 1

$$\Delta T_a = \frac{\sqrt{T_{ref}\left(T_{ref} + T_{add} + 4T_n\right) + 2T_n^2 - T_a\left(T_a + T_{add} - 2T_{ref}\right)}}{\sqrt{2df\tau R}}.$$

(21)

The radiometer measurement range dT_a with concerned input block is equal T_{add}. This fact has been taken into account in equation (21).

From (21) follows that the sensitivity depends on the specific antenna signal and is variable in measurement range. In [13] , the attention was taken into the fluctuating sensitivity dependence on antenna signal. But the case of a large antenna signal was regarded in this article. Owing to super noiseless amplifier with noise floor of tens of Kelvin creation whole radiometric system self-noise were considerably reduced, which were comparable with measured signals.

Usable sensitivity occurs for the antenna signal $T_a = T_{ref} - T_{add}/2$ in the middle of measurement range

$$\Delta T_{a,max} = \frac{\sqrt{2\left(T_{ref} + T_n\right)^2 + \dfrac{T_{add}^2}{4}}}{\sqrt{2 df \tau R}}.$$

(22)

To achieve the necessary threshold of sensitivity during the designing of radiometer with the input block obtain the relation for product calculation from (22)

$$\tau R = \frac{\left(T_{ref} + T_n\right)^2 + \dfrac{T_{add}^2}{8}}{df \Delta T_{a,max}^2}.$$

(23)

Consider t and R determination strategy by example. Let it be required to secure the measurement dynamic range 0...300 K for a receiver with noise temperature $T_n = 200$ K and receiving band df = 100 MHz. It is required to provide a minimum signal-detection threshold 0.05 K ($\Delta T_{a,max} = 0.05$ K) in this range. The modulating frequency in radiometer is 1 kHz ($t_{mod} = 500$ ms).

Firstly determine the levels of reference signals $T_{ref} = T_{a,max} = 300$ K, $T_{add} = T_{ref} - T_{a,min} = 300$ K.

Then using (23) find $\tau R = \dfrac{\left(T_{ref} + T_n\right)^2 + \dfrac{T_{add}^2}{8}}{df \Delta T_{a,max}^2} = 1.045$.

Choose t = $30t_{mod}$ = 0.015 s to ensure the necessary dynamic properties of the regulatory system. Whence R = 1.045/t = 69. Then, for the modulation period 1 ms, the measurement time will be 69 ms.

To provide the necessary sensitivity the required capacity of radiometer output digital code is determined by further calculations. Find the number of minimum antenna signal values on the measurement range N = $dT_a/DT_{a,max}$ = 6000 (consider that for a given input block structure a measurement range $dT_a = T_{add}$). Then a rates number of radiometer output digital code n = $\log_2 N$ = $\log_2 6000$. Get code size n = 13 by rounding up the whole.

Using similar calculations for other input blocks in Figure 2(b) and Figure 2(c), it is useful to get the necessary data for a modified zero-radiometer designing.

As example of sensitivity calculations by Formula (29) determine $\sqrt{2df\tau R}$ which is situated in its denominator. $\sqrt{2df\tau R}$ characterizes radiometer receiver and signal processing at low frequency. Choose radiometer

parameters as more typical: df = 100 MHz, modulation frequency 1 kHz, time constant of low frequency filter t =30ms, R = 1000 which correspond to signal time storage equal to 1 second. $\sqrt{2df\tau R}$ is dimensionless quantity equal to 78383.7. The graph in Figure 3 shows calculated by Equation (21) threshold sensitivity dependence of the modified radiometer for the most typical in remote sensing measurement ranges for a receiver with noise temperature T_n = 50 K. These calculations imply that the sensitivity within the range of measurements remains almost invariant. At the edges of measurement range the sensitivity takes the same minimum value and insignificantly increases in the middle of the range. The sensitivity depends on the upper limit of measurement range. The minimum threshold ΔT_a increases in the case of range expansion in the direction of higher measurement antenna temperatures measurement.

Found by similar way: transfer characteristic, measurement range, fluctuation sensitivity of radiometer with the input blocks are tabulated in Table 2 and their schemes are shown in Figure 2(b). Calculation of minimum detectable signal for both blocks gives the same results with the same measurement ranges.

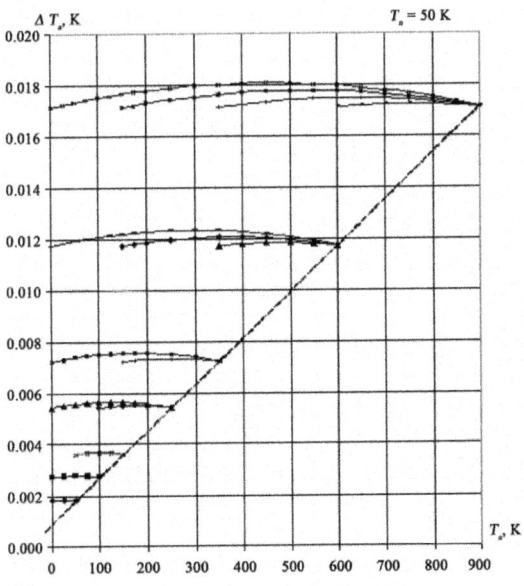

Figure 3. Fluctuation sensitivity responses of the radiometer with input block (Figure 2(a)) from antenna signal of different measuring range.

The dependence of fluctuation sensitivity on antenna signal diagrams for various full scales receivers with different noise temperatures are plotted in Figure 4. The sensitivity does not remain the same in the measurement range and changes during the antenna signal modifying with almost linear variation. ΔT_a reaches a peak at the maximum antenna signal and trough at the antenna signal equal to minimum value of the measurement range.

For all considered input units T_n augment leads to a proportional increase of the minimum detectable antenna signal and sensitivity deterioration occurs linearly with noise temperature increasing of the receiver and the input path.

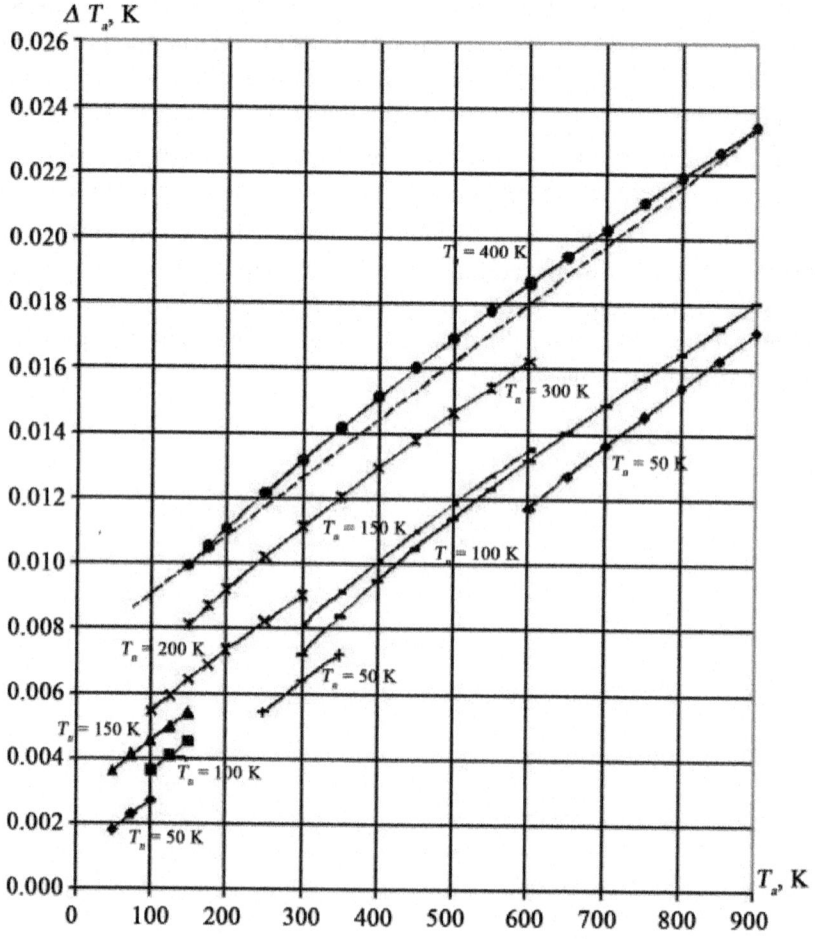

Figure 4. Fluctuation sensitivity responses of the radiometer with input block (Figure 2(b) and Figure 2(c)) from antenna signal of different measuring range and different receiver noise characteristics.

Table 2: The transfer characteristics, the measurement ranges, the radiometer sensitivity with the input blocks shown in Figure 2(b) and Figure 2(c)

Input block	The transfer characteristics, the measurement ranges, the radiometer sensitivity
Input block by Figure 2(b) scheme	$t_{puz} = \dfrac{T_a - T_{ref}}{T_{add}} \times t_{mod}$; $T_a = T_{ref} + T_{add}\dfrac{t_{puz}}{t_{mod}}$ $T_{a,min} = T_{ref}\ (t_{puz}=0)$; $T_{a,max} = T_{ref} + T_{add}\ (t_{puz}=t_{mod})$ $dT_a = T_{a,max} - T_{a,min} = T_{add}$ $\Delta T_a = \dfrac{\sqrt{T_a\left(2T_{ref}+T_{add}+T_a+4T_n\right)+2T_n^2-T_{ref}\left(T_{ref}+T_{add}\right)}}{\sqrt{2df\tau R}}$; $\Delta T_{a,max} = \dfrac{\sqrt{2\left(T_{ref}+T_n\right)^2+2T_{add}^2+4T_{add}\left(T_{ref}+T_n\right)}}{\sqrt{2df\tau R}}$ $\tau R = \dfrac{\left(T_{ref}+T_n\right)^2+T_{add}^2+2T_{add}\left(T_{ref}+T_n\right)}{df\Delta T_{a,max}^2}$
Input block by Figure 2(c) scheme	$t_{puz} = \dfrac{T_a - T_{ref}}{T_{add}-T_{ref}} \times t_{mod}$. $T_a = T_{ref} + \left(T_{add}-T_{ref}\right)\times\dfrac{t_{puz}}{t_{mod}}$. $T_{a,min} = T_{ref}$ $T_{a,max} = T_{add}$; $dT_a = T_{add} - T_{ref}$ $\Delta T_a = \dfrac{\sqrt{T_a\left(T_{add}+T_{ref}+T_n+4T_n\right)+2T_n^2-T_{add}T_{ref}}}{\sqrt{2df\tau R}}$. $\Delta T_{a,max} = \dfrac{T_{add}+T_n}{\sqrt{df\tau R}}$; $\tau R = \dfrac{\left(T_{add}+T_n\right)^2}{df\Delta T_{a,min}^2}$

EXPERIMENTAL RESEARCHES

The formulas obtained in section 5 for sensitivity calculations were tested in experimental researches. Sensitivity determine experiments were carried out for the considered input blocks with the radiometer at the wavelength 6.5 sm (total noise temperature of the whole system is equal to 600 K, receiving bandwidth—100 MHz, modulation frequency—1 kHz). Antenna signal changing was performed by antenna tilt variation. Directional diagram covered an area in which there were no sources of synthetic electromagnetic radiation. The sensitivity was determined for different antenna signals.

Radiometer was reconstructed and 8 series of 16 measurements with 5 minutes intervals were performed. Storage time was set equal to 1 second. Average values were calculated for each series. These values were plotted on a graph shown in Figure 5. The values of the minimum detectable antenna signal were imaged on ordinate axis; the values of the measured antenna signal were imaged on abscissa axis. Curves 1 and 2 on the graphs are constructed according to the obtained analytical dependences of the radiometer with the input blocks sensitivity calculation. Schemes of these blocks are shown in Figure 2(a) and Figure 2(c), respectively. Experimentally obtained values of sensitivity are represented by vertical lines on the given graphs. The scatter of the sensitivity measured values over standard periods of time at a constant antenna signal was taken into account. Maximum spread between findings obtained by theoretical and experimental way was about 20% with standard deviation of 8% ... 10%.

CONCLUSIONS

On the basis of the combined pulse modulation and the original principle of signal processing, the algorithm of the following system functioning was developed. This algorithm showed that it was useful to carry out the auto-zero balance in the radiometer by changing of pulse-width signal duration. According to this algorithm, at the output of the radiometer after constant component exclusion in the first half-period of rectangular symmetrical modulation pulse volt-second areas, equalization is performed.

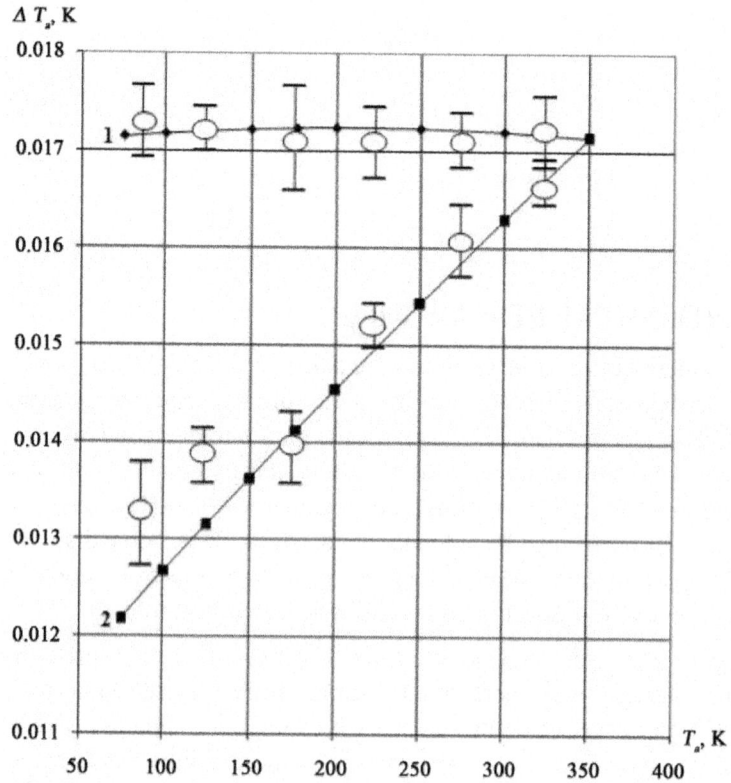

Figure 5: Radiometer fluctuation sensitivity with different receiving blocks at the input.

This procedure is equivalent to the signal energies' equalization at the radiometer receiver input at the different modulation half-periods. Zero voltage in the second modulation half period is an indicator of pulse volt- second areas' equality. As a result, a mathematical model is found. This model establishes a linear relation between antenna effective temperature and duration of reference noise signal modulated by pulse-width-law.

The analysis of the fluctuation sensitivity of this measuring method pointed to variable sensitivity, the dependence on the measured antenna signal. Formulas for fluctuation sensitivity calculation were obtained and experimental verification was carried out for three proposed schemes of the modified radiometer input blocks with different measured signals' range.

ACKNOWLEDGEMENTS

This work was supported by Russian Foundation for Basic Research, grants No. 13-07-98009, No. 15-07-04971.

REFERENCES

1. Sharkov, E.A. (2003) Passive Microwave Remote Sensing of the Earth: Physical Foundations. Springer/PRAXIS, Berlin.

2. Astafieva, N.M., Raev, M.D. and Sharkov, E.A. (2006) Portret Zemli iz kosmosa. Globalnoe radioteplovoe pole. Priroda, 9, 75-86.

3. Dicke, R.H. (1946) The Measurement of Thermal Radiation at Microwave Frequencies. Review of Scientific Instruments, 17, 268-275. http://dx.doi. org/10.1063/1.1770483

4. Orhaur, T. and Waltman, W. (1962) Switched Load Radiometer. Public National Radio Astronomy Observatory, 1, 179-204.

5. Ryle, M. and Vonberg, D.D. (1948) An Investigation of Radio-Frequency Radiation from the Sun. Proceeding of the Royal Society, 193, 98-119. http://dx.doi.org/10.1098/rspa.1948.0036

6. Troitskiy, V.S., Lubina, A.G. and Zolotov, A.V. (1951) Sravnenie teplovih shumov nekotorix materialov nulevim metodom. Dokladi Akademii Nauk SSSR, 4, 583-586.

7. Hardy, W.N., Gray, K.W. and Love, A.W. (1974) An S-Band Radiometer Design with High Absolute Precision. IEEE Transactions on Microwave Theory and Techniques, MTT-22, 382-391.

8. Lawrence, R.F., Harrington, R.F. and Higdon, N.S. (1982) Flight Test Evaluation of a Noise Injection Dicke Microwave Radiometer Employing Digital Signal Processing. IEEE MTT-S International Microwave Symposium Digest, New York, 15-17 June 1982, 90-92.

9. de Maagt, P.J.I., Oerlemans, R.A.E., van Gestel, J.C.A.M. and Herben, M.H.B.J. (1992) A Novel Radiometer Receiver Stabilization Method. International Journal of Infrared and Millimeter Waves, 13, 1075-1097. http://dx.doi.org/10.1007/BF01009052

10. Brown, S.T., Desai, S., Wenwen, Lu. and Tanner, A.B. (2007) On the Long-Term Stability of Microwave Radiometers Using Noise Diodes for Calibration. IEEE Transactions on Geoscience and Remote Sensing, 45, 1908-1920.

11. Vlaby, F.T., Moore, R.K. and Fung, A.K. (1981) Microwave Remote Sensing. Artech House, Norwood.

12. Sironi, G., Inzani, P., Limon, M. and Marchioni, C. (1990) Evaluation of Small Signals with a Differential Radiometer (with Application to Radio Observations at 2.5 Ghz). Measurement Science and Technology, 1, 1119-1121. http://dx.doi.org/10.1088/0957-0233/1/10/025

13. Esepkina, N.A., Korolkov, D.V. and Pariyskiy, U.N. (1973) Radioteleskopi i radiometri. Nauka, Moscow, 415 p.

Chapter 13

MONITORING INSTRUMENTATION IN UNDERGROUND STRUCTURES

Alireza Maghsoudi, Behzad Kalantari

Department of Civil Engineering, University of Hormozgan, Bandar Abbas, Iran

ABSTRACT

Nowadays underground structures are very important. Based on observations of engineering; properties during geotechnical construction are an integral part of the design of underground structures. This research presents instrumentation as a tool to assist with these measurement observations, determine the need for modifications to loading or support arrangement. Also apart from above construction control, instrumentation is also indispensable for site investigation, design verification and safety of the structure. Instrumentation used in the construction of tunnels and subways can be implemented in three stages before, during operation and during operation are examined. Metro Railway Tunnels are constructed in populated area and have a more comprehensive instrumentation and monitoring program that additionally includes monitoring of ground conditions, underground water levels, tilt and settlement of nearby buildings or other structures of interest in the vicinity of the tunnel alignment. Instrumentation monitoring for metro railway tunnels includes monitoring of the structures under construction together with the ground, buildings and other facilities within the predicted zone of influence. Furthermore, instrumentation and subway tunnels in and around them increase accuracy of the different layers of the earth and excavation of the surrounding structures and make safety and accuracy. This paper presents the features of sophisticated instrumentation available today for geotechnical monitoring. A wide range of sophistic have been described with their applications ted electronic and mechanical instrumentation with different instrumentation schemes used to meet the requirements of different types of structures.

INTRODUCTION

Based on stability and strength of the subway tunnel and by instrument stability is a measure of confidence in the estimates. The initial cost of the used instruments, always look for a way to get the most information with the least utility. Deformation Measurements Instruments are installed in the tunnel roof and at selected points along the tunnel walls to monitor vertical, horizontal, and longitudinal in tunnel direction deformation components. The number of points and their detailed location depends on the size of the tunnel and the excavation sequencing in multiple drift applications. Also, as a minimum of the wall for each drift including temporary elements should be equipped with a device capable of measuring deformations [1] .

The designer of geotechnical construction works with naturally occurring materials, and does not find their exact engineering properties. He may carry out tests in the laboratory on the samples picked up from the field, and sometimes change the naturally occurring materials to make them more suitable for his needs. Instrumentation is used to measure the response (deformation, stress and etc.) of soil or rock to changes in loading or support arrangements, and from the measurements taken, the need for modifications to the loading or support arrangements is determined. This illustrates the basic reason why instrumentation is generally of immense value during geotechnical construction [2] .

FUNDAMENTALS OF INSTRUMENTATION

A good instrumentation program should have one or more of the following purposes in mind such as site investigation which instruments are used to characterize and determine initial site conditions. Common parameters of interest in a site investigation are pore pressure, permeability of soil, slope stability etc. In other side, design verification based on instruments are used to verify design assumptions. Instrumentation data from the initial stage of a project may show the need or provide the opportunity to modify the design in later stages. In this process Construction Control are instruments which installed to monitor the effects of construction. Instrument data helps the engineer to determine how fast construction can proceed without adverse effects on the foundation soil and construction materials used.

Safety of instruments can provide early warning of impending failure. In case of metro railway tunnels instruments provide early warning through real time monitoring systems available on the internet for any excessive and undue ground movements affecting the adjoining premises, structure and utilities like the railways, power lines, water lines etc. In this side, legal protection based on instruments provide designers and contractors the basis of a legal defence should resident and owners of adjacent properties blame construction for damage to their property and life. This aspect gains prominence in constructions in populated areas such as for underground metro railways.

Instrument performance is used to monitor the in-service performance of a structure. For example, monitoring leakage, pore water pressure and deformation can provide an indication of the performance of a dam. Monitoring loads on rock bolts and movements within a tunnel can provide an indication of the stability of tunnel [3] [4] .

TEST METHODS FOR UNDERGROUND STRUCTURES

In Situ Stress

In situ tests are used to directly obtain field measurements of useful soil and rock engineering properties. In soil, in situ testing include both index type tests, such as the Standard Penetration Test (SPT) and tests that determine the physical properties of the ground, such as shear strength from cone penetration Tests (CPT) and ground deformation properties from pressure meter tests (PMT). One significant property of interest in rock is it's in situ stress condition. Horizontal stresses of geological origin are often locked within the rock masses, resulted in a stress ratio (K) often higher than the number predicted by elastic theory.

Hydraulic Fracturing

Typically conducted in vertical boreholes. A short segment of the hole is sealed off using a straddle packer. This is followed by the pressurization by pumping in water. The pressure is raised until the rock surrounding the hole fails in tension at a critical pressure.

Flat Jack Test

Like all other stress measuring methods, the flat jack method is also based on certain assumptions such as 1) the stress at the test site is compressive, 2) the surrounding rock mass is homogeneous, elastic and isotropic, 3) the rock deforms symmetrically around the slot, 4) the state of stresses at the measurement site is uniform, and 5) the stress applied to the rock by the flat jack is uniform. The method can be used only in the walls of the underground opening after the opening has been made, and measures the induced stresses from which the in situ stresses are calculated. Flat jack, the Following breakdown, the shut-in pressure, and the lowest test-interval pressure at which the hydrocrack closes completely under the action of the stress acting normal to the hydro fractures. In a vertical test hole the hydro fractures are expected to be formed in vertical and perpendicular to the minimum horizontal stress) Figure 1.

Some Common Mistakes form-meter, reference pins, a standard distance bar and a hydraulic pump with pressure gauges are the main items of equipment required for the test. Flat jack is inserted into the slot, cemented in place, and pressurized.

When the pins have been returned to the initial separation, the pressure in jack approximates the initial stress normal to the jack. The slot cut in the horizontal direction yields induced stress tangential to the boundary of the opening and the slot cut in the vertical direction yields induced stress, parallel to the axis of the opening, at the respective slot locations (Figure 2) [6] .

In actual tests, the length of the slot may be bigger than the jack, and the slot may not be of fixed width. Further, the stress acting in the plane parallel to the major axis of the slot also affects the contraction of the slot. The formulae given below account for the effect of all the above and can be used for evaluating the stresses (Equations (1) and (2)) [7] .

$$PK_1 = P_\theta K_2 + P_H K_3 \text{(For jack position horizontal)} \qquad (1)$$

$$PK_1 = P_\theta K_3 + P_H K_2 \text{(For jack position vertical)} \qquad (2)$$

MODULUS OF DEFORMATION

Modulus of deformation is recognized as one of the important parameters governing the rock mass behavior. It is experienced that tunnel sections generally take long time to stabilize. These time-dependent deformation effect the tunnel lining along with the modulus of deformation of the rock mass. The study suggests that time for ninety per cent tunnel closure and deformation is less both for good quality rock and larger span of tunnel. The concept of retarded creep appears to be valid for weak and dry rock masses around a supported tunnel in non-squeezing condition. The important various tests include [8] :

Pressure Meter Test (PMT)

Where P is the flat jack cancellation pressure, $P\theta$ is the induced stress tangential to the boundary of the opening and P_H is the induced stress parallel to the axis of the opening. K1, K2 and K3 are constants which are determined from parameters such as the length of the jack, slot dimensions, distance between reference pins, etc. Induced stresses Pθ and PH can be determined by solving simultaneous Equations (1) and (2). At each test site, several tests should be carried out to obtain statistically viable results. Typical stress-deformation plots from flat jack test data for very good quality basalt and poor quality volcanic breccia rock mass from Koyna H.E. Project, Maharashtra, with EM values of 62.8 GPa and 4.80 GPa, respectively, are shown in (Figure 3).

This test is performed with a cylindrical probe placed at the desired depth in a borehole. The Menard type pressure meter requires pre-drilling of the borehole; the self-boring type pressure meter advances the whole itself, thus reducing soil disturbance. The PENCEL pressure meter can be set in place by pressing it to the test depth or by direct driving from ground surface or from within a predrilled borehole. The hollow center PENCEL probe can be used in series with the static cone penetrometer.

The pressure required expanding the cylindrical membrane to a certain volume or radial strain is recorded in a borehole. It is applicable for soft rocks. Pressure meter are two types of High pressure instrument and Low pressure instrument (Figure 4) [9] .

The pressure meter test is an in-situ testing method which is commonly used to achieve a quick and easy measure of the in-situ stress-strain relationship of the soil which provides parameters such as the elastic modulus gives typical tests results (Figure 5).

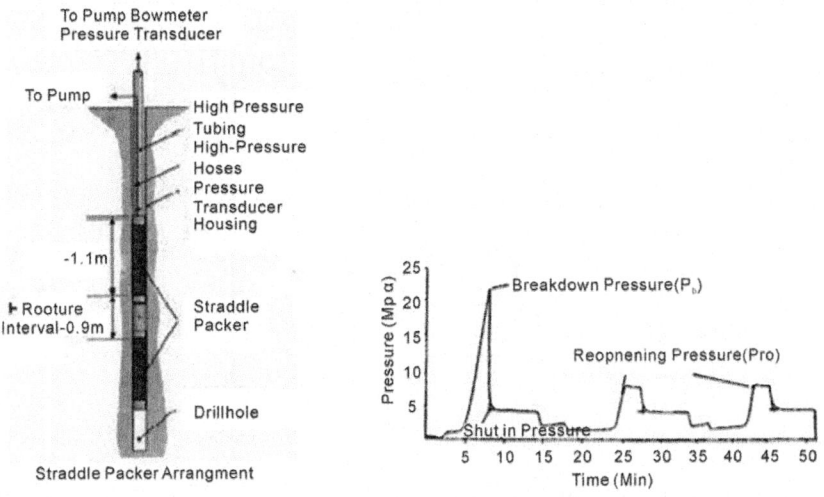

Figure 1: Schematic diagram and graphs of typical hydraulic fracturing system.

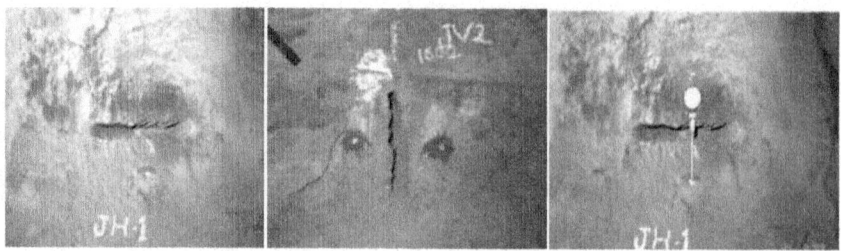

Figure 2: Slots made by jack hammer drilling and convergence measurement by de-form-meter.

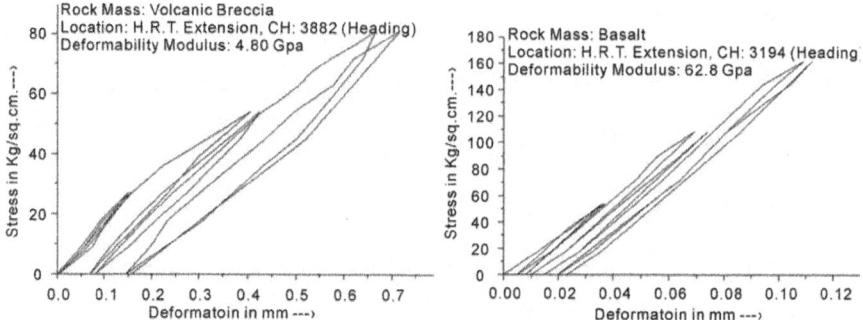

Figure 3: Typical stress verses deformation envelopes from flat jack tests.

From the pressure strain curves, the untrained shear strength, the soil shear modulus and the in-situ horizontal stress can be obtained. On completion of a test, the cone is advanced to the next test depth [10] [11] .

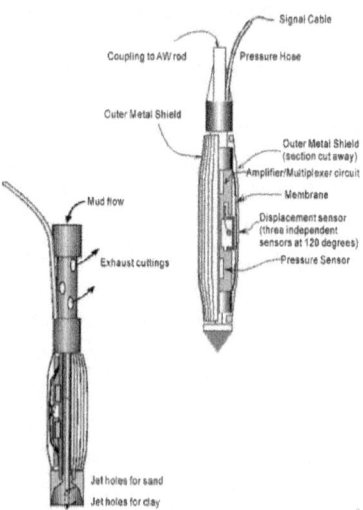

Figure 4: Low pressure instrument and high pressure instrument.

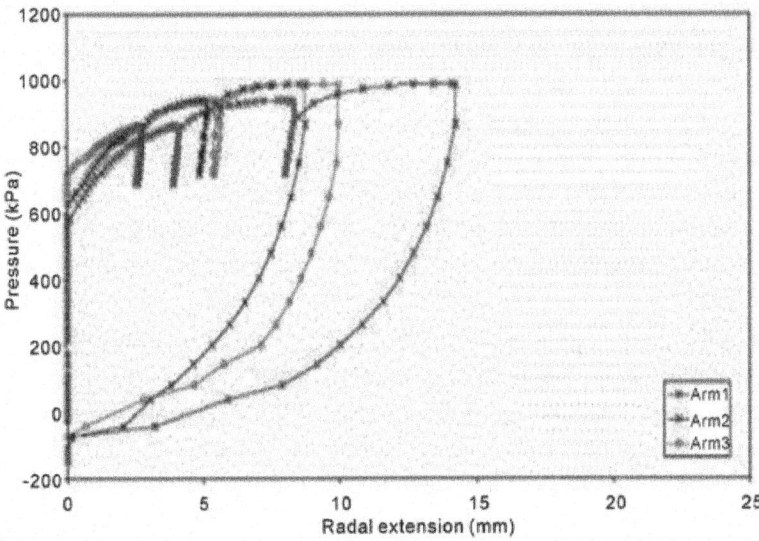

Figure 5: Typical results for a cone pressure meter test.

Figures and Tables

In addition the pressure meter membrane can be inflated to a large radial strain, approximately 50%, to ensure that the pressure meter limit pressure is approached at full inflation. The speed of cone penetration testing enables the cone pressure meter to be an economical alternative to conventional self-boring and Menard-type pressure meter tests.

Plate Bearing Test

A relatively flat rock surface us sculptured and level with mortar to receive circular bearing plates 20 to 40 inches in diameter. Loading a rock surface and monitoring the resulting displacement. This is easily arranged in the underground gallery. The site may be selected carefully to exclude loose, highly fractured rock. The modulus of subgrade reaction is calculated from the relation [12] .

$$K = \frac{P}{0.125} \left(\frac{Kg/cm^2}{cm} \right)$$

MEASURING DISPLACEMENTS

The objectives of ground deformation monitoring are different in mountain and urban tunnels. In mountain tunnels, the main objective of deformation measurements during construction is to ensure that ground pressures are adequately controlled, i.e., there exists an adequate margin of safety against collapse, including roof collapse, bottom heave, failure of the excavation face, yielding of the support system, etc. Adequate control of ground pressures ensures a safe and economical structure, well adapted to the inherent heterogeneity of ground conditions. As a result of these differences in objectives, design philosophies, and construction techniques, the types and required accuracy of the measured ground deformations vary between the two classes of tunnels, as follows[13] :

One ft2 Plate is placed on a soil surfacee. A load P is placed on the plate and the settlementis measured as a function of P. Two different types of shapes can be developed depending on the density of the soil. From this, we can get a peak and residual bearing capacity. Notice the shape of the curves; they look a lot like the curves seen in a direct shear test. This is because the direct shear test reflects the conditions at failure (Figure 6).

1)   In mountain tunnels, considerable ground deformations are deliberately permitted (and often provoked) in order to reduce the initially very large "geostatic" loads on the temporary support by increasing ground de-confinement.

2)   In urban tunnels, the main objective is limiting ground deformations around the tunnel and thus causing the minimum possible movement and disturbance at ground surface and the structures founded there.

The important various tests include extensometer. Extensometers can be installed in tunnel to measure the displacement of rock mass surrounding the tunnel. From the measured data of extensometers, some information of the rock mass, for example, the initial stresses and modulus of deformation, can be obtained. The displacements of extensometer are often induced by many inadequate factors, such as poor grouting at anchor point of extensometer [14] .

In a tunnel in Tehran is obtained using a extensometer. Station 10 is a full station in which convergence meter, extensometer and load cell installed that useful information can be obtained from this station such as: determination of plastic zone radius, determination of effective drilling zone around tunnel, stability and evaluation of tunnel using direct strain control technique.

Thus, some limitations of the measured data should be clarified before the measured data of extensometer are applied for further study. Figure 7 shows two types of instruments for measuring the vertical settlement of a series of locations in the ground.

Drawing of characteristic curve of the earth and determination of appropriate support system for tunnel. In figure 8 the actual rock mass displacement charts around the tunnel in the left and right walls are shown [15] .

DETERMINATION OF PORE PRESSURE

The soil is incorporated in the design of tunnels by calculating the soil pressures that are exerted on the tunnel and pressures necessary to have a stable tunnel face.

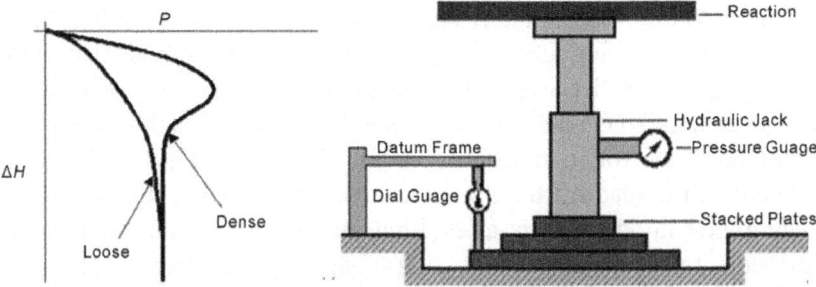

Figure 6: Plate bearing test (PBT) and plate bearing test results for loose and dense soils.

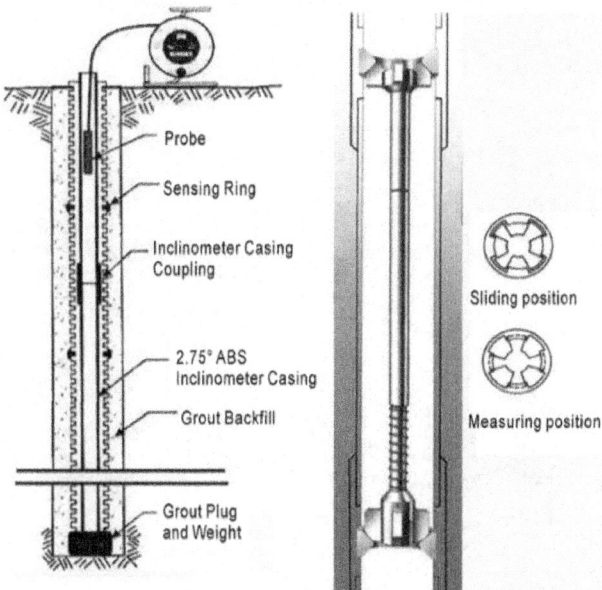

Figure 7: Magnetic extensometer and fixed re-installable micrometer (FIM) of the sliding micrometer.

Pore pressures are taken into account, but generally only a hydrostatic pressure distribution is assumed. However, the construction of a tunnel uses liquids pressurized with pressures different from the hydrostatic pressure: betonies slurry at the tunnel face and grout mortar at the tail void. This will lead to excess pore pressures. Knowledge of these excess pore pressures appears to be of importance for the stability of the tunnel face and the final pressures on the lining. Piezometer is the simplest form of Piezometric measurement is the monitoring of water levels in cased boreholes (internal diameter of casing 20 to 60 mm) [16] .

The plot below shows the results from a typical test in the height of the water level in the rising tube (piezometer) is measured with an electric water level gauge. The device is mounted on the end of the casing and the gauge body lowered into the tube. Once the electrode integrated in the plumb touches the surface of the water, an electric circuit is closed and a pilot lamp comes on. The depth is read off the scale on the cable. The height of the tube end is determined in advance by a levelling measurement. Pour water pressure measurement was carried out by various types of piezometers[17] :

Casagrande Piezometer: A simple open standpipe placed in a borehole, it consists of a PVC standpipe with a permeable membrane tip (Figure 9).

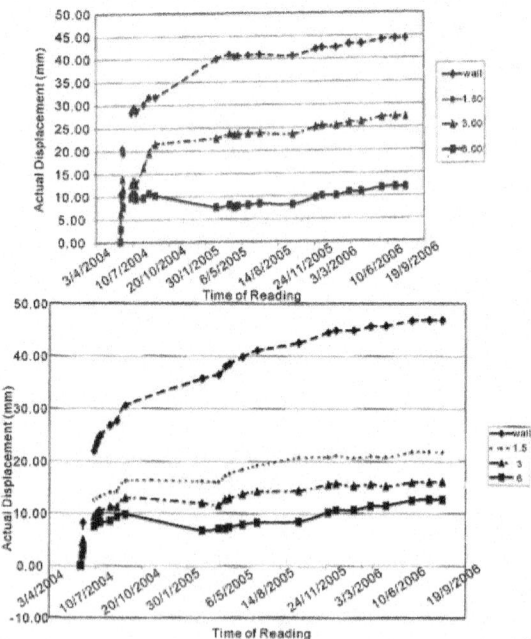

Figure 8: The actual displacement curve of the right rock mass and left rock mass in station 10 of Taloun pilot tunnel.

Pneumatic Piezometer: Dry nitrogen is introduced down an inlet tube which is blocked with a flexible diaphragm (Figure 9).

The first series and the second series of tests were run for 280 days (Figure 10).

GEOPHYSICAL TESTING METHODS

Geophysical tests are indirect methods of exploration in which changes in certain physical characteristics such as magnetism, density, electrical resistivity, elasticity, or a combination of these are used as an aid in developing subsurface information. Geophysical methods provide an expeditious and economical means of supplementing information obtained by direct exploratory methods, such as borings, test pits and in situ testing; identifying local anomalies that might not be identified by other methods of exploration; and defining strata boundaries between widely spaced borings for more realistic prediction of subsurface profiles. The important various tests include [18] :

Seismic refraction tomography (SRT) which is important to investigate the loosing rock zone of tunnel surface for anchoring design of the some deformed rock mass. Conventional refraction inversion methods use a layered model approach. The subsurface is divided into a number of continuous constant velocity layers having velocities and thicknesses which are modified through interactive forward modeling in an effort to match travel times determined from the field data. These conventional methods require sections of travel time curves to be mapped to refractors, a task that can be difficult in several geological situations.

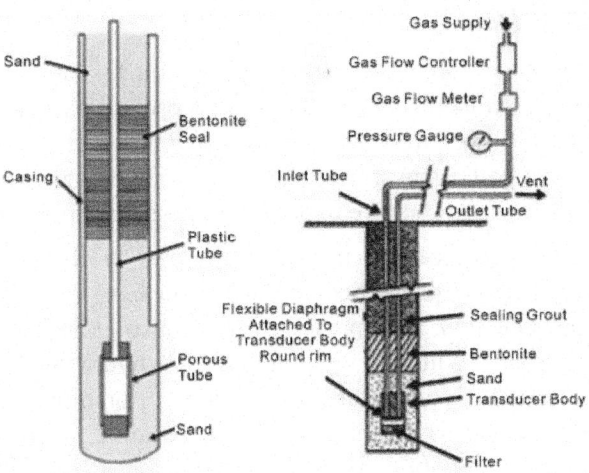

Figure 9: Casagrande Piezometer and Pneumatic Piezometer.

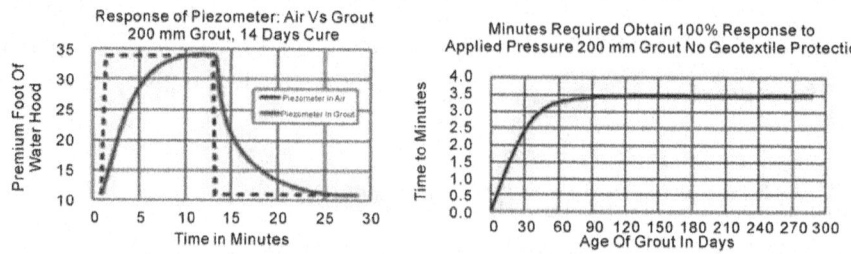

Figure 10: Casagrande piezometer and pneumatic piezometer.

Detectors (geophones) are positioned on the ground surface at increasing distance from a seis mic impulse source, also at the ground surface. The time required for the seismic impulse to reach each geophone is recorded a seismic device like that shown in figure 11 shows contour lines that represented boundaries between compression wave velocity gradients [18] .

The cross-hole seismic technique was used to identify the extent of the Excavation Damage Zone (EDZ) is in a test tunnel and in a tunnel under construction. At least 2 boreholes are required: a source borehole within which a seismic pulse is generated, and a receiver borehole in which a geophone records generated compression and shear waves (figure 12). For increased accuracy additional receiver boreholes are used. Receivers must be properly oriented and securely in contact with the side of the borehole [19] . Boreholes deeper than about 30 ft should be surveyed using an inclinometer or other device to determine the travel distance between holes [20] , [21] .

CONCLUSION

Based on mentioned topic, laboratory testing is a fundamental element of a geotechnical investigation. The ultimate purpose of laboratory testing is to utilize repeatable procedures to refine the visual observations and field testing conducted as part of the subsurface field exploration program, and to determine how the soil or rock will behave under the imposed conditions.

Laboratory testing of soil and rock samples to characterize materials; determine engineering properties for use in design; testing in accordance with specified standards; high-pressure 3-axial testing, large diameter 3-axial tests, dynamic testing with high-response, electro-hydraulic, closed loop MTS equipment, and high temperature testing; slake durability testing. Detailed soil laboratory testing is required to obtain accurate information including classification, characteristics, stiffness, strength, etc. for design and modeling purposes. Testing are performed on selected representative samples (disturbed and undisturbed) in accordance with ASTM standards. Rock Testing is generally controlled by the discontinuities within the rock mass and not the properties of the intact material. Standard rock testing evaluate physical properties of the rock included density and mineralogy in thin-section analysis. The mechanical properties of the intact rock core included uniaxial compressive strength, tensile strength, static and dynamic elastic constants, hardness of indices.

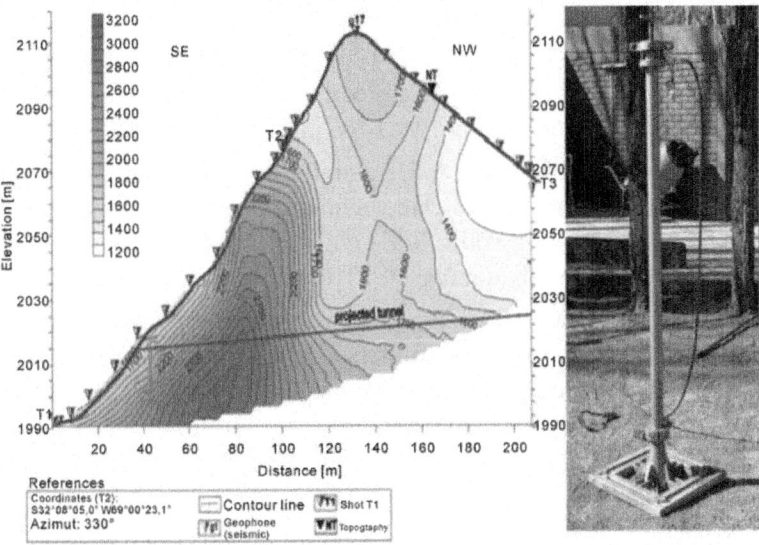

Figure 11: TAMPER seismic wave generation system and seismic refraction tomography profile.

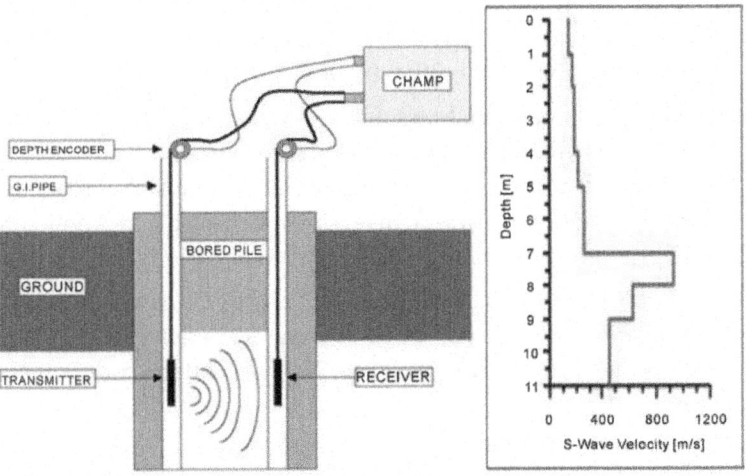

Figure 12: Cross-Hole devices and Resulting S-wave velocity profile.

REFERENCES

1. Millogo, Y., Morel, J.-C., Traoré, K. and Ouedraogo R. (2012) Microstructure, Geotechnical and Mechanical Characteristics of Quicklime-Lateritic Gravels Mixtures Used in Road Construction. Construction and Building Materials, 26, 663-669.http://dx.doi.org/10.1016/j.conbuildmat.2011.06.069

2. Sikora, Z. and Ossowski, R. (2013) Geotechnical Aspects of Dike Construction Using Soil-Ash Composites. Procedia Engineering, 57, 1029-1035.http://dx.doi.org/10.1016/j.proeng.2013.04.130

3. Argyroudis, S., Kaynia, A.M. and Pitilakis, K. (2013) Development of Fragility Functions for Geotechnical Constructions: Application to Cantilever Retaining Walls. Soil Dynamics and Earthquake Engineering, 50, 106-116.http://dx.doi.org/10.1016/j.soildyn.2013.02.014

4. Cabalar, A.F., Cevik, A. and Gokceoglu, C. (2012) Some Applications of Adaptive Neuro-Fuzzy Inference System (ANFIS) in Geotechnical Engineering. Computers and Geotechnics, 40, 14-33. http://dx.doi.org/10.1016/j.compgeo.2011.09.008

5. Kolat, C., Ulusay, R. and Suzen, M.L. (2012) Development of Geotechnical Microzonation Model for Yenisehir (Bursa, Turkey) Located at a Seismically Active Region. Engineering Geology, 127, 36-53. http://dx.doi.org/10.1016/j.enggeo.2011.12.014

6. Kim, H.-S., Cho, G.-C., Lee, J.Y. and Kim, S.-J. (2013) Geotechnical and Geophysical Properties of Deep Marine Fine-Grained Sediments Recovered during the Second Ulleung Basin Gas Hydrate Expedition, East Sea, Korea. Marine and Petroleum Geology, 47, 56-65.http://dx.doi.org/10.1016/j.marpetgeo.2013.05.009

7. Shaaban, F., Ismail, A., Massoud, U., Mesbah, H., Lethy, A. and Abbas, A.M. (2013) Geotechnical Assessment of Ground Conditions around a Tilted Building in Cairo-Egypt Using Geophysical Approaches. Journal of the Association of Arab Universities for Basic and Applied Sciences, 13, 63-72. http://dx.doi.org/10.1016/j.jaubas.2012.06.002

8. Jia, S.P., Zhao, Y.Q. and Zou, C.S. (2012) Numerical Solution to Identification Problems of Material Parameters in Geotechnical Engineering. Procedia Engineering, 28, 61-65.http://dx.doi.org/10.1016/j.proeng.2012.01.683

9. Papaioannou, I. and Straub, D. (2012) Reliability Updating in Geotechnical Engineering Including Spatial Variability of Soil. Computers and Geotechnics, 42, 44-51.http://dx.doi.org/10.1016/j.compgeo.2011.12.004

10. Ghorbani, M., Sharifzadeh, M., Yasrobi, S. and Daiyan, M. (2012) Geotechnical, Structural and Geodetic Measurements for Conventional Tunnelling Hazards in Urban Areas—The Case of Niayesh Road Tunnel Project. Tunnelling and Underground Space Technology, 31, 1-8. http://dx.doi.org/10.1016/j.tust.2012.02.009

11. Al-Mukhtar, M., Khattab, S. and Alcover, J.-F. (2012) Microstructure and Geotechnical Properties of Lime-Treated Expansive Clayey Soil. Engineering Geology, 139-140, 17-27.http://dx.doi.org/10.1016/j.enggeo.2012.04.004

12. Ching, J.Y. and Phoon, K.-K. (2013) Quantile Value Method versus Design Value Method for Calibration of Reliability-Based Geotechnical Codes. Structural Safety, 44, 47-58.http://dx.doi.org/10.1016/j.strusafe.2013.04.003

13. Ghayoomi, M., Dashti, S. and McCartney, J.S. (2013) Performance of a Transparent Flexible Shear Beam Container for Geotechnical Centrifuge Modeling of Dynamic Problems. Soil Dynamics and Earthquake Engineering, 53, 230-239.http://dx.doi.org/10.1016/j.soildyn.2013.07.007

14. Juang, C.H. and Wang, L. (2013) Reliability-Based Robust Geotechnical Design of Spread Foundations Using Multi-Objective Genetic Algorithm. Computers and Geotechnics, 48, 96-106. http://dx.doi.org/10.1016/j.compgeo.2012.10.003

15. Ma, L., Luo, H.B. and Chen, H.R. (2013) Safety Risk Analysis Based on a Geotechnical Instrumentation Data Warehouse in Metro Tunnel Project. Automation in Construction, 34, 75-84. http://dx.doi.org/10.1016/j.autcon.2012.10.009

16. Wu, X.Z. (2013) Trivariate Analysis of soil Ranking-Correlated Characteristics and Its Application to Probabilistic Stability Assessments in Geotechnical Engineering Problems. Soils and Foundations, 53, 540-556. http://dx.doi.org/10.1016/j.autcon.2012.10.009

17. Katzenbach, R., Leppla, S., Vogler, M., Seip, M. and Kurze, S. (2013) Soil-Structure-Interaction of Tunnels and Superstructures during Construction and Service Time. Procedia Engineering, 57, 35-44. http://dx.doi.org/10.1016/j.proeng.2013.04.007

18. Gaaver, K.E. (2012) Geotechnical Properties of Egyptian Collapsible Soils. Alexandria Engineering Journal, 51, 205-210. http://dx.doi.org/10.1016/j.aej.2012.05.002

19. Abela, J.M., Potts, D.M., Vollum, R.L. and Izzuddin, B.A. (2013) Geotechnical Analysis of Blinding Struts in Cut-And-Cover Excavations.

Computers and Geotechnics, 48, 179-191.http://dx.doi.org/10.1016/j. compgeo.2012.07.007

20. Heerten, G. (2012) Reduction of Climate-Damaging Gases in Geotechnical Engineering Practice Using Geosynthetics. Geotextiles and Geomembranes, 30, 43-49.http://dx.doi.org/10.1016/j. geotexmem.2011.01.006

21. Raptakis, D.G. (2012) Pre-Loading Effect on Dynamic Soil Properties: Seismic Methods and Their Efficiency in Geotechnical Aspects. Soil Dynamics and Earthquake Engineering, 34, 69-77. http://dx.doi. org/10.1016/j.soildyn.2011.09.003

Chapter 14

THE FUNDAMENTAL OPERATING PRINCIPLES OF ELECTRONIC ROOT CANAL LENGTH MEASUREMENT DEVICES

M. H. Nekoofar[1,2], M. M. Ghandi[3] , S. J. Hayes[2] & P. M. H. Dummer[2]

[1]Department of Endodontics, Faculty of Dentistry, Tehran University of Medical Science, Tehran, Iran

[2]Division of Adult Dental Health, School of Dentistry, Cardiff University, Cardiff, UK; and

[3]Department of Electronic Systems Engineering, University of Essex, Colchester, UK

ABSTRACT

Nekoofar MH, Ghandi MM, Hayes SJ, Dummer PMH. The fundamental operating principles of electronic root canal length measurement devices. International Endodontic Journal, 39, 595–609, 2006.

It is generally accepted that root canal treatment procedures should be confined within the root canal system. To achieve this objective the canal terminus must be detected accurately during canal preparation and precise control of working length during the process must be maintained. Several techniques have been used for determining the apical canal terminus including electronic methods. However, the fundamental electronic operating principles and classification of the electronic devices used in this method are often unknown and a matter of controversy. The basic assumption with all electronic length measuring devices is that human tissues have certain characteristics that can be modelled by a combination of electrical components. Therefore, by measuring the electrical properties of the model, such as resistance and impedance, it should be possible to detect the canal terminus. The root canal system is surrounded by dentine and cementum that are insulators to electrical current. At the minor apical foramen, however, there is a small hole in which conductive materials within the canal space (tissue, fluid) are electrically connected to the periodontal ligament that is itself a conductor of electric current. Thus, dentine, along with tissue and fluid inside the canal, forms a resistor, the value of which depends on their dimensions, and their inherent resistivity. When an endodontic file penetrates inside the canal and approaches the minor apical foramen, the resistance between the endodontic file and the foramen decreases, because the effective length of the resistive material (dentine, tissue, fluid) decreases. As well as resistive properties, the structure of the tooth root has capacitive characteristics. Therefore, various electronic methods have been developed that use a variety of other principles to detect the canal terminus. Whilst the simplest devices measure resistance, other devices measure impedance using either high frequency, two frequencies, or multiple frequencies. In addition, some systems use low frequency oscillation and/or a voltage gradient method to detect the canal terminus. The aim of this review was to clarify the fundamental operating principles of the different types of electronic systems that claim to measure canal length.

INTRODUCTION

The presence of bacteria and their by-products within the root canal system predisposes to apical periodontitis. In essence, treatment of a microbial inflammatory disease is directed at the elimination of the antigenic source through the elimination of infection (Chugal et al. 2003). Root canal treatment involves removing microorganisms from within the pulp space, and the filling of the root canal system to prevent reinfection. Furthermore, restoring the tooth to prevent recontamination and reinfection is essential (Heling et al. 2002). In other words, the goal of root canal treatment is to control infection through the debridement, disinfection and filling of the root canal system (Lin et al. 2005). It is generally accepted that root canal treatment procedures should be limited to within the root canal system (Ricucci 1998). To attain this objective the endpoint of the root canal system, the canal terminus, should be detected as precisely as possible during preparation of the canal. Therefore, one of the main concerns in root canal treatment is to determine how far instruments should be advanced within the root canal and at what point the preparation and filling should terminate (Katz et al. 1991).

MORPHOLOGY OF THE ROOT CANAL TERMINUS

Kuttler (1955) concluded that a root canal had two main sections, a longer conical section in the coronal region consisting of dentine and a shorter funnelshaped section consisting of cementum located in the apical portion. The shape of this apical portion is considered to be an inverted cone (Fig. 1); its base being located at the major apical foramen. The apex of the inverted cone is the minor foramen that is often thought to coincide with the apical constriction regarded as being at or near the cemento-dentinal junction (CDJ) (Kuttler 1958). In other words, the most apical portion of the root canal system narrows from the opening of the major foramen, which is within cementum, to a constriction (minor foramen) before widening out in the main canal to produce an hour-glass shape (Fig. 1).

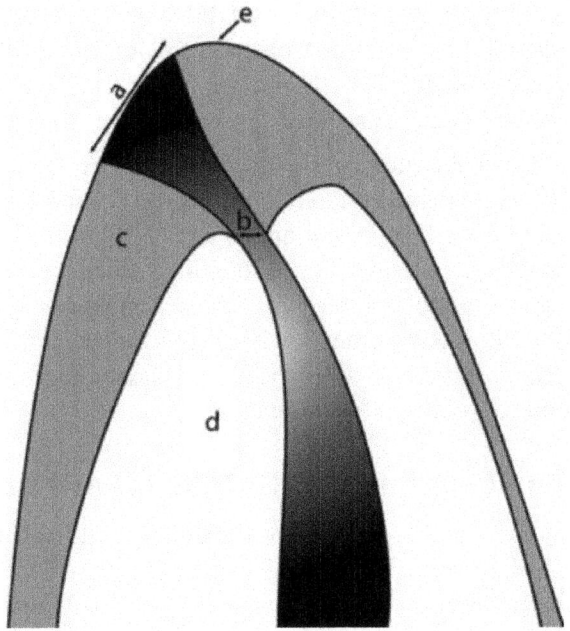

Figure 1: Idealized anatomy of apical portion of root (a) major apical foramen, (b) minor apical foramen (apical constriction) that may be coincident with the cemento-dentinal junction (CDJ), (c) cementum, (d) dentine and (e) root apex.

It is well known that the major apical foramen is not a uniform shape but can be asymmetrical (BlaskovicSubat et al. 1992). Furthermore, its position on the root tip varies. For example, Stein & Corcoran (1990) reported that with increasing age the deviation of the major foramen from the root tip increased, whilst others have reported that the frequency of the deviation depended on the type of teeth (Blaskovic-Subat et al. 1992). Moreover, deviation of the foramen can occur as a result of pathological changes, the most common being external root resorption (Malueg et al. 1996).

The root canal terminus is considered by many to be the CDJ (Kuttler 1955, 1958, Ricucci 1998, Ponce & Fernandez 2003). In some instances the CDJ coincides with the pulp and periodontal tissue junction, where the pulp tissue changes into apical periodontal tissue (Seltzer 1988). Theoretically, the

CDJ is the appropriate apical limit for root canal treatment as at this point the area of contact between the periradicular tissues and root canal filling material is likely to be minimal and the wound smallest (Palmer et al. 1971, Seltzer 1988, Katz et al. 1991, Ricucci & Langeland 1998). The term 'theoretically' is applied here because the CDJ is a histological site and it can only be detected in extracted teeth following sectioning; in the clinical situation it is impossible to identify its position. In addition, the CDJ is not a constant or consistent feature, for example, the extension of the cementum into the root canal can vary (Ponce & Fernandez 2003). Therefore, it is not an ideal landmark to use clinically as the end-point for root canal preparation and filling.

Defining the canal terminus as the apical constriction and not the CDJ is also problematical, as the topography of the apical constriction is not constant (Dummer et al. 1984). Indeed, the apical constriction can have a variety of morphological variations that makes its identification unpredictable. In clinical practice, the minor apical foramen is a more consistent anatomical feature (Katz et al. 1991, Ponce & Fernandez 2003) that can be regarded as being the narrowest portion of the canal system and thus the preferred landmark for the apical end-point for root canal treatment.

DETERMINING THE ROOT CANAL TERMINUS

Various techniques have been used for determining the position of the canal terminus and thus measure the working length of root canals. The most popular method has been the use of radiographs. However, although it is generally accepted that the minor apical foramen and apical constriction is on average located 0.5–1.0 mm short of the radiographic apex (Katz et al. 1991, Morfis et al. 1994) there are wide variations in the relationship of these landmark that would result in under- or over-preparation of canals with an obvious impact on the position of the root filling (Stein & Corcoran 1990, Olson et al. 1991). Thus, many studies have shown that canal lengths determined radiographically vary from actual root canal lengths by a considerable amount (Kuttler 1955, 1958, Green 1956, Green 1960, Dummer et al. 1984, Forsberg 1987a,b, Martinez-Lozano et al. 2001).

The accuracy of radiographic methods of length determination depend on the radiographic technique that has been used (Forsberg 1987a, Katz et al. 1991). For example, Sheaffer et al. (2003) revealed that higher density radiographs were more desirable for measuring working length. Forsberg (1987b) reported that tooth length determined by the bisecting angle technique, either correctly or incorrectly angulated, was less accurate than the paralleling technique. Although radiographs are a critical and an integral part of endodontic therapy

(Vertucci 2005) there is an ongoing need to reduce exposure to ionizing radiation whenever possible (Brunton et al. 2002, Pendlebury et al. 2004).

It is well known that the major apical foramen is not always located at the radiographic apex of the root; rather, it often lies on the lingual/buccal or mesial/ distal aspect. If the major foramen deviates in the lingual/buccal plane (Fig. 2), it is difficult to locate its position using radiographs alone, even with multiplane angles (Schaeffer et al. 2005).

One of the innovations in root canal treatment has been the development and production of electronic devices (McDonald 1992) for detecting the canal terminus. Their functionality is based on the fact that the electrical conductivity of the tissues surrounding the apex of the root is greater than the conductivity inside the root canal system provided the canal is either dry or filled with a nonconductive fluid (Custer 1918). Suzuki (1942) indicated that the electrical resistance between a root canal instrument inserted into a canal and an electrode applied to the oral mucous membrane registered consistent values. Based on these findings, Sunada (1962) reported that when the tip of an endodontic instrument had reached the periodontal membrane through the 'apical foramen', the electrical resistance between the instrument and the oral mucous membrane was a constant value. Based on this fundamental principle, these resistance-based devices should be able to detect the periodontal tissue at the 'apical foramen'. Clearly, they do not assess the position of the root apex and the name 'electronic apex locator' is not appropriate; 'electronic apical foramen locator' or 'electronic root canal length measurement device' (ERCLMD) as a generic name would be more appropriate.

The manufacturers of more recent electronic devices claim their products locate the apical constriction (http://www.vdw-dental.com/home_e/index. html, http://www.jmoritausa.com/Marketing/pdf/RootZX_IFU. pdf, http:// www.averon.ru/english/dental/equipment/ apex.htm, http://www.micro-mega. com/anglais/produits/ apexpointer/index.php, http://www.parkell.com/master. html, http://www.parkell.com/foramatron.html). Their claims are based on the fact that these newer devices operate using different electronic principles compared with the original resistance-based devices (Kim & Lee 2004). However, the evidence suggests their claims are not correct, for example, Hoer & Attin (2004) reported that accurate determination of the apical constriction was only successful in 51–64% of canals depending on the device used. The probability of determining the position between the minor and major foramen was between 81 and 82% of cases. Welk et al. (2003) also reported that the ability of various types of ERCLMDs to determine the 'minor diameter' was between 90.7 and 34.4%.

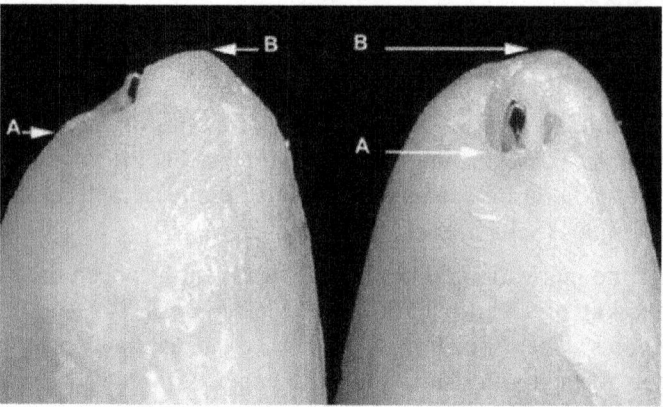

Figure 2: (a) Major apical foramen (apical opening) with protruding instruments; (b) root apex.

Because of the hazards of radiation (Katz et al. 1991, Brunton et al. 2002, Pendlebury et al. 2004), the technical problems associated with radiographic techniques (Heling & Karmon 1976, Forsberg 1987a) and to avoid over-instrumentation beyond the canal terminus (ElAyouti et al. 2002) electronic working length determination has gained popularity amongst both general dentists and endodontists (Frank & Torabinejad 1993). The electronic method is also more convenient to the patient and has potential to enable root canal treatment to be performed during pregnancy (Trope et al. 1985). Unfortunately, most manufacturers do not define the exact nature of their devices nor how they operate electronically. Classifying and describing the devices by 'Generation' is not helpful to clinicians and is better suited to marketing issues. In essence, it is not possible to classify all the various products on the market; rather, only those whose fundamental operating principles have been released by the manufacturer can be categorized (Table 1). Clearly, with the limited information provided by manufacturers the classification of electronic devices used to measure canal length is a matter of controversy and ignorance (Nekoofar 2005).

The aim of this article is to describe the fundamental operating principles of ERCLMDs and classify them on this basis (Table 1). Initially, a review of basic electronics is presented in order to provide an understanding of electronic devices and circuits; the electronic function of ERCLMDs is then discussed.

Atom Structure

To understand the basics of electronics, the structure of the atom, which is the smallest particle of materials that retain their characteristics, should be defined. Atoms are made of electrons, protons and neutrons. According to the classic Bohr model (Coombs 1999); atoms have a planetary type of structure that comprises a central nucleus surrounded by orbiting electrons. The nucleus consists of positively charged particles called protons and uncharged particles called neutrons. The basic particles of negative charge are called electrons.

Electrons orbit the nucleus of an atom at certain distances from the nucleus. Electrons that are in orbits further from the nucleus are less tightly bound than those closer to the nucleus. This is because the force of attraction between the positively charged nucleus and the negatively charged electron decreases with increasing distance from the nucleus. Therefore, electrons existing in the outermost shell of an atom are relatively loosely bound to the atom. For example, in the copper atom the most outer shell has one electron and when that electron gains sufficient thermal energy it can break away from the parent atom and become a free electron. In copper at room temperature, a large number of these free electrons are present that are not bound to any atom and are free to move. Free electrons make copper an excellent conductor and make electrical current possible. Other conductive materials may have similar characteristics but with different conductivity determined by their atom structure.

Table 1: Categorization of electronic root canal length measurement devices

Type	Name	Manufacturer
Resistance-based ERCLMDs	Endodontic Meter® Endometer® Faramatron 4® Apex Finder®	Parkell Inc., New York, NY, USA
Low frequency oscillation	Sono-Explorer® Sono-Explorer Mark II®	Hayashi Dental Supply, Tokyo, Japan
High frequency devices (capacitance-based devices)	Endocater®	Hygenic Corp., Akron, OH, USA
Capacitance and resistance look-up table	Elements®Diagnostic unit	SybronEndo, Orange, CA, USA
Voltage gradient (difference in impedance with three nodes)	No commercial model available	
Two frequencies, impedance difference	Apit® Apex pointer® Root ZX®	Osada, Tokyo, Japan MicroMega, Besançon, France J. Morita Co., Kyoto, Japan
Impedance ratio (Quotient)	Justy II Endy 5000	Parkell Inc., New York, NY, USA Parkell Inc., New York, NY, USA
Multifrequency	Endo Analyzer® (8005) AFA Apex Finder® (7005)	SybronEndo, Orange, CA, USA SybronEndo, Orange, CA, USA
Unknown	Foramatron® D10	

Ions and Electrolyte

When the number of electrons changes in an atom, the electrical charge will change. If an atom gains electrons, it picks up an imbalance of negatively charged particles and will become negative. If an atom loses electrons, the balance between positive and negative charges is shifted in the opposite direction and the atom will become positive. In either case, the magnitude (+1, +2,)1,)2, etc.) of the electrical charge will correspond to the number of electrons gained or lost. Atoms that carry electrical charges are called ions (regardless of whether they are positive or negative). A cation is an ion that has lost electrons and acquired a positive charge; an anion is an ion that has gained electrons and acquired a negative charge.

Not only do electrons flow along a wire in an electric circuit, but electrons can also be carried through water if it contains ions in solution. Ionic solutions that conduct electricity in a manner similar to wire are called electrolytes. The conductance of electrolytes is the result of the movement of ions through the solution towards the electrodes. When two electrodes in a solution are part of a complete electrical circuit, the cations (+) are attracted to the negative pole (cathode) and the anions (()) are attracted to the positive pole (anode).

The conductivity of any particular ion will be affected by the ease with which the ion can move throughout the water. The ease with which any ion moves through a solution depends on factors such as the total charge and the size of the ion; large ions offer greater resistance to motion through the electrolyte than small ions. The greater the number of ions present, the greater the electrical conductivity of the solution.

Electrical Charge, Voltage and Current

Electrical charge, symbolized by Q, is either positive or negative. The electron is the smallest particle that exhibits negative electrical charge. When an excess of electrons exists in a material, there is a net negative electrical charge, and conversely, a deficiency of electrons forms a net positive electrical charge. Materials with charges of opposite polarity are attracted to each other and materials with charges of similar polarity are repelled.

A certain amount of energy must be used in the form of work to overcome the forces and move the charges a given distance apart. All opposite charges possess a certain potential energy because of the separation between them. The difference in potential energy of the charges is voltage. Voltage, symbolized by V, is the driving force in electric circuits and is what establishes current. The unit of voltage is the volt. Voltage provides energy to electrons or ions that allows them to move through a circuit. This movement is electrical current characterized by I, which results in work being done in an electric circuit. The measurement unit of current is ampere.

Resistance

When there is a current of free electrons in a material the electrons occasionally collide with atoms. These collisions cause the electrons to lose some of their energy, and thus restrict their movement. The more collisions, the more the flow of electrons is restricted. This restriction varies with the type of material, the property of which is called resistance and is designated as R; it is expressed in the unit of ohms (X). However, when electrical current is formed by ions, the current is restricted by other means. When a voltage (potential difference) is applied between two points in an electrolyte, ions between them will be attracted by the opposite charge and so will move between the points and produce a current. The resistance of such electrolytic solutions depends on the concentration of the ions and also on the nature of the ions present, in particular, their charges and mobilities. Thus, resistance is a variable that depends on concentration.

This physical effect is called resistivity which is represented by q. For each material, q can be a constant value at a given temperature. Thus, the resistance of an object can simply depend on three factors: (i) resistivity, (ii) length and (iii) cross-sectional area. The formula for the resistance of an object of length l and cross-sectional area A is:

$$R = \frac{\rho \times l}{A}.$$

(1)

The formula shows that resistance increases with resistivity and length, and decreases with cross-sectional area. In fact, resistivity (q) is the parameter that classifies conductive materials from insulators. Insulators cannot conduct electric currents because all their electrons are tightly bound to their atoms. A perfect insulator would allow no charge to be forced through it, however, no such substance is known at room temperature. The best insulators offer high but not infinite resistance at room temperature. For instance, the resistivity of human thorax bone at normal temperatures is approximately 16 000 Ω m-[1] (Geddes & Baker 1967) whilst for blood it is 100 times less at approximately 162 Ω m-[1] (Rush et al. 1963). Thus, bone is a relatively poor conductor whilst blood is a relatively good conductor of electric current.

Electric Circuits and the Human Body

Current through the body, not voltage, is the cause of an electrical shock. When a point on the body comes in contact with a voltage and another point comes in contact with different voltage, there will be a current through the body from one point to the other. The severity of the resulting electrical shock depends on the amount of voltage and the path that the current takes through the body (Bridges 2002). To measure the effects of current on the human body, the amount of current should be calculated. This is dependent on the potential difference, the impedance driving the potential difference and the resistance within the body between the points of contact (Niple et al. 2004). The human body has resistance that depends on many factors, including body mass, skin moisture and points of contact of the body with a voltage potential. The human body does not sense currents less than a few milliamperes (Gandhi 2002). However, about hundred milliamperes of current will cause fatal damage; especially if it is connected for more than several seconds (Bridges 2002).

Ohm's Law

Ohm's law describes the mathematical relationship between voltage, current and resistance in a circuit. Ohm determined that if the voltage across a resistor is increased, the current through the resistor will increase and conversely, if the voltage is decreased the current will decrease. Ohm's law also shows that if the voltage (V) is kept constant, less resistance (R) results in more current (I), and more resistance results in less current. Ohm's law can be stated as follows:

$$I = \frac{V}{R} \quad \text{or} \quad V = I \times R \quad \text{or} \quad R = \frac{V}{I}.$$

(2)

Electrolyte solutions also obey Ohm's law just as metallic conductors do. From a macroscopic point of view, ionic conduction of solutions is similar to electron conduction through solid objects. In the latter, electrons are moving without ion cores, whilst in the former, charges are moving as ions. Although water itself is a poor conductor of electricity, the presence of ions in solution decreases the resistance considerably. The resistance of such electrolytic solutions depends on the concentration of the ions and also on the nature and size of the ions present.

Direct Current and Alternating Current

Direct current (DC) is a fixed amount of current per unit of time, whereas the amount of an alternating current (AC) alternates with time. The sine wave (or sinusoidal wave) is the fundamental type of alternating current and alternating voltage. Figure 3 is a graph showing the general shape of a sine wave, which can be either an alternating current or an alternating voltage. In fact, when a sinusoidal voltage source is applied to a resistive circuit, it will result in an alternating sinusoidal current.

The current (or voltage) varies with time, starting at zero, increases to a positive maximum, returns to zero, and then increases to a negative maximum before again returning to zero, thus completing a full cycle. The time (in seconds) required for a given sine wave to complete one full cycle is called the period (T).

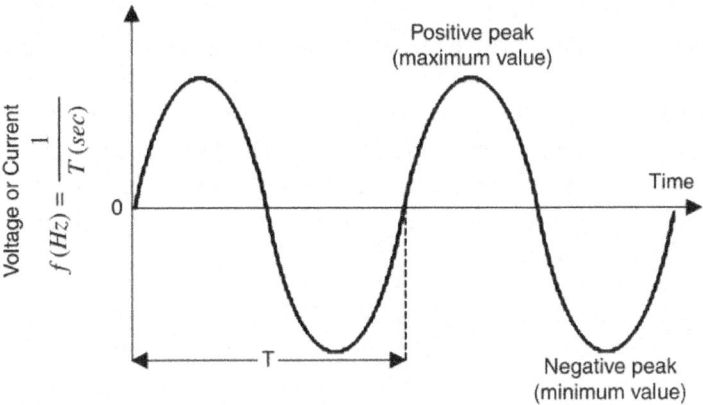

Figure 3: A sine wave as an alternating voltage or current.

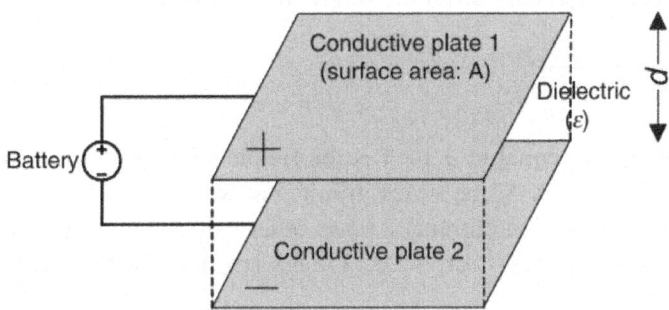

Figure 4: A simple capacitor connected to a battery (DC voltage source).

Frequency is the number of cycles that a sine wave completes in 1 s. The more cycles completed in 1 s, the higher the frequency. Frequency is symbolized by f and is measured in units of hertz (Hz).

Capacitor

A structure of two conductive materials with an insulator between them forms an electrical device called a capacitor. In its simplest form shown in Fig. 4, a capacitor is constructed of two parallel metal plates separated by an insulating material called a dielectric. When a capacitor is connected to a DC voltage source, electrons (negative charge) move from one plate to another, making one plate acquire a negative charge and the other a positive charge. When the voltage source is disconnected, the capacitor would still retain the stored charge and a voltage will remain across it. The amount of charge that a capacitor can store will determine its capacitance.

The following parameters are important in establishing the capacitance of a capacitor: plate area (A), plate separation (d) and dielectric constant (ε). A large plate area produces a large capacitance and a smaller plate area produces a smaller capacitance. Conversely, the plate separation (d) is inversely proportional to the capacitance, i.e. a greater separation of the plates reduces the capacitance. Finally, the insulating material between the plates (the dielectric) will directly influence the capacity by its dielectric constant (e) as shown in the equation:

$$C = \frac{\epsilon \times A}{d}$$

(3)

As a result of the insulator, a capacitor will block constant DC. However, it allows the AC to pass with an amount of opposition that depends on its capacitance and the frequency of the AC. This opposition is called capacitive reactance (X_c) calculated from the following formula:

$$X_C = \frac{1}{2\pi \times f \times C}$$

(4)

where π is almost equal to 3.14, f is the frequency and C is the capacitance. When f is zero (DC), X_C becomes infinite and so blocks the DC. At nonzero frequencies (alternating current) it takes other values and becomes analogous to the resistance of a resistor, hence, Ohm's law applies to capacitive circuits as follows:

$$I = \frac{V}{X_c} \quad \text{or} \quad V = I \times X_c \quad \text{or} \quad X_C = \frac{V}{I}.$$

(5)

Impedance and its Measurement

In a circuit that has both capacitors and resistors, the total amount of opposition to an alternating current is called impedance which is represented by Z. Again, Ohm's law applies to these circuits:

$$I = \frac{V}{Z} \quad \text{or} \quad V = I \times Z \quad \text{or} \quad Z = \frac{V}{I}.$$

(6)

The value of impedance in a circuit that has both resistors and capacitors depends on the resistance values (R) of its resistors and the reactance (X_C) values of its capacitors.

There are several methods to measure the impedance value of a material. The basic method is to apply an electrical current to the material and measure the resulting voltage. According to Ohm's law (equation 6), the division product of the voltage value over the current value gives the value of impedance. If the material comprises resistive elements only, a DC can be enough for this measurement (using equation 2). However, in the presence of capacitive elements, an AC current highlights the capacitive characteristics of the impedance as well as the resistive part. The frequency of the AC current will influence the measured impedance value as the capacitive component of the impedance is variable with frequency (see equation 4).

The use of DC is impractical for measurement of the resistance of an electrolyte, as the electrodes become polarized. The behaviour of electrolytic solutions is of huge technological importance as well as of great scientific interest. However, during electroconductivity measurement, polarization can be prevented by using a high-frequency alternating current, so that the quantity of electricity carried during one half-cycle is insufficient to produce any measurable polarization, although various electrolytes in different conditions may exhibit different conductivities.

It should be added that an impedance, in electrical terms, has two properties: amplitude (or simply value) and phase. For simplicity in the context of this paper, impedance is normally identified with its value. In addition, as mentioned above, there are a number of methods to measure impedance value or phase electrically, e.g. Wheatstone bridge (Lazrak et al. 1997), Phase-sensitive detection (Hartov et al. 2000), and Sine-wave correlation (Van Driessche et al. 1999). Electronic publications will provide more details (Coombs 1999).

Electrical Features of Tooth Structure

Root canals are surrounded by dentine and cementum that are insulators to electric current. At the minor apical foramen, however, there is a small hole in which conductive materials within the canal are electrically connected to the periodontal ligament that is a conductor of electric current. The resistive material of the canal (dentine, tissue, fluid) with a particular resistivity forms a resistor, the value of which depends on the length, cross-sectional area and the resistivity of the materials (Fig. 5). If an endodontic file penetrates inside the canal, and approaches the canal terminus, the resistance between the end of the instrument and the apical portion of the canal decreases, because the effective length of the resistive material inside the canal (l in Fig. 5) decreases.

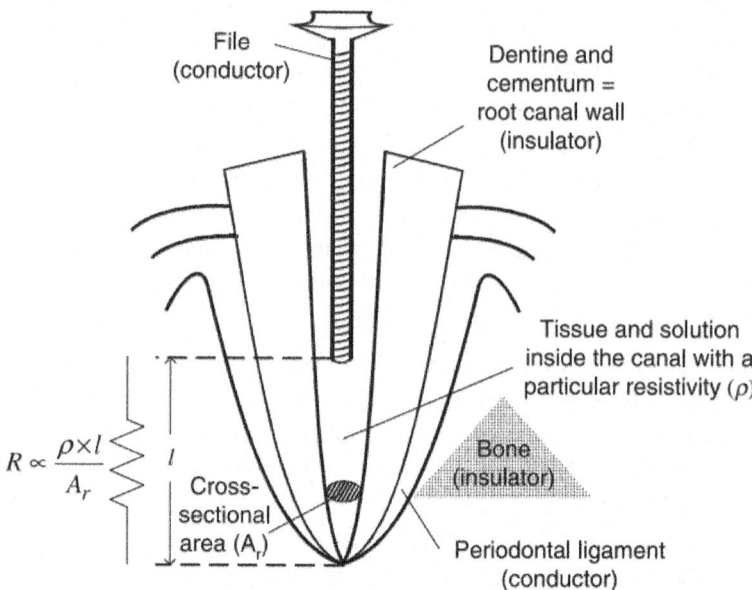

Figure 5: The structure of the tooth during root canal treatment in terms of electrical conductivity, and the resistance of the model.

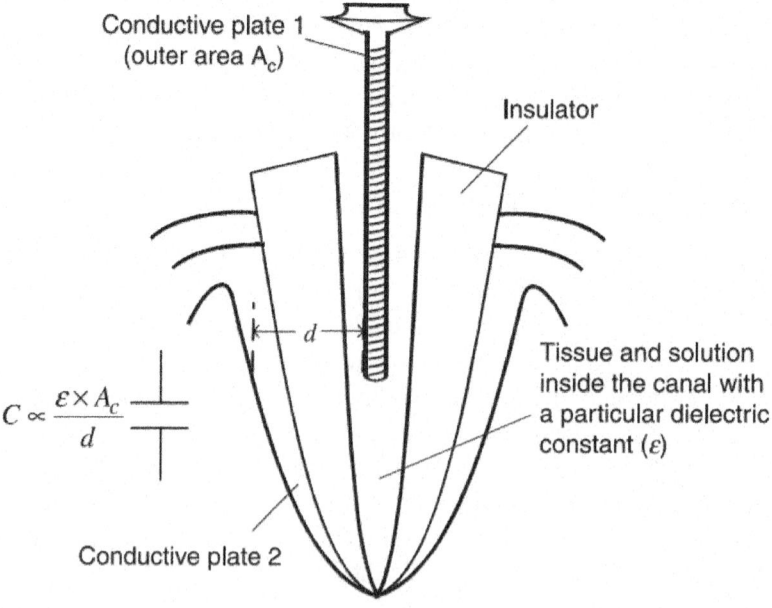

Figure 6: The capacitance of the tooth during root canal treatment.

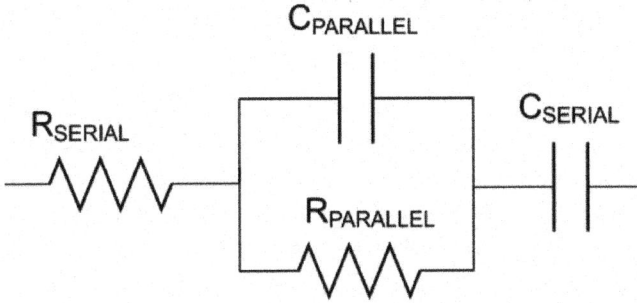

Figure 7: The simplified electronic model of a tooth proposed by Meredith & Gulabivala (1997).

As well as resistive properties, the structure of the tooth has capacitive characteristics. Assume the file, with a specific surface area, to be one side of a capacitor and the conductive material (e.g. periodontal ligament) outside the dentine being the other plate of that capacitor. Tissue and fluid inside the canal, in addition to the cementum and dentine of the canal wall, can be considered as separators of the two conductive plates and determine a dielectric constant e. This structure forms a capacitor, much more complex than symbolized in Fig. 6.

The electrical structure of the canal is much more complicated than the resistive and capacitive elements described above and the exact modelling of it is not a straightforward task (Meredith & Gulabivala 1997). Meredith & Gulabivala (1997) proposed an equivalent circuit that modelled the root canal system including periapical tissues. They found that the root canal acted as a complex electrical network with resistive and capacitive elements. It exhibited complex impedance characteristics having series and parallel resistive and capacitive components with a simplified model shown in Fig. 7.

ELECTRONIC ROOT CANAL LENGTH MEASUREMENT DEVICES

The fundamental assumption of ERCLMDs is that human tissues have certain characteristics that can be modelled by means of a combination of electrical components. Therefore, by measuring the electrical properties (e.g. resistance, impedance) of that equivalent electric circuit, some clinical properties (such as the position of a file) can be extracted. Custer (1918) introduced a new electrical approach for locating the canal terminus that was dependent on the fact that the electrical conductivity of the tissues surrounding the apex of the root is greater than the conductivity inside the root canal system, coronal to the canal terminus. Custer (1918) noted that this difference in conductivity values could be detected more easily if the canal was dry or filled with a nonconductive liquid such as alcohol.

In other words, he discovered that the electrical resistance, the inverse value of conductivity, near the 'foramen' was much less than in the coronal region of the root canal. Therefore, Custer (1918) located the position of the 'foramen' by applying a voltage between the 'alveolus opposite the root apex' and the 'broach inside the pulp' and measuring the value of the electrical current (with a 'milammeter'). In his pioneer experiment, using the technology of that time, Custer's electrical circuit had three 'dry cells': a 'milammeter', a negative and a positive electrode.

When the circuit was connected, a small positive voltage was applied to the fine insulated 'broach' which was introduced into the 'pulp' and slowly penetrated inside it. When the 'broach' approached the 'foramen', as a result of a significant increase in the electrical conductivity, the electrical current became more and as a consequence a 'certain movement in the index finger' of the ammeter was observed. Custer (1918) concluded that this movement, which was proportional to the electrical current and so the electrical conductivity, would be a reliable guide to the position of the broach relative to the 'apical foramen'. Subsequently, Suzuki (1942) in his experimental study on iontophoresis in dog's teeth indicated that the electrical resistance between a root canal instrument inserted into a canal and an electrode applied to the oral mucous membrane registered consistent values. Based on Suzuki's finding, Sunada (1962) reported that a specific value of the resistance would determine the position of the root canal terminus. He determined that when the tip of an endodontic instrument had reached the periodontal membrane through the 'apical foramen', the electrical resistance between the instrument and the oral mucous membrane was approximately equal to 6.5 k Ω. He also claimed that if the reamer perforated the canal wall or floor of the pulp chamber and reached the periodontal membrane, the electrical resistance between the mucous membrane and the perforated periodontal membrane was almost equal to the resistance shown at the apex. In his first in vivo experiment, Sunada (1962) used a simple microammeter to measure the length of 71 canals. The micro-ammeter had two electrodes, one attached to the oral mucous membrane, the other to an endodontic instrument positioned in the root canal; the resistance was measured when the tip of the endodontic instrument was situated at the apex (the length of the teeth were already known 'by means of a measuring wire and X-ray') the resistance of the periodontal membrane was calculated by dividing the voltage by the value of current that the device measured.

In other words, dentine, enamel and cementum are electrical insulators, soft tissue, including the periodontal ligament, is a conductor. The device established a circuit in the mouth that originated in the device, ran through an endodontic file via the attached probe, and extended down the canal out of the foramen and into the periodontal ligament. The circuit continued through the patient's mucosa and eventually completed the loop by running into the lip clip that was connected to the device through a return wire. Sunada (1962) also reported that the patient's age, type or shape of teeth and the diameter of the canal had no influence on the results. The mean value of the resistance of the circuit between the canal terminus and the lip clip was 6.5 k Ω.

Resistance-based ERCLMDs

Resistance-based ERCLMDs relied on the assumption that the circuit between the endodontic file and the lip clip could be modelled by a simple resistive circuit (Fig. 8). Therefore, a small DC was applied to that circuit and the voltage was measured. By dividing the value of the voltage by the value of the current, the resistance value of the circuit was calculated.

Many electronic root canal length measuring devices have since been marketed (Table 1) using the same principles. The differences between them are basically in the design of the electrical circuits and in the mode of their display. However, they can all be classified as 'resistance-based' ERCLMDs.

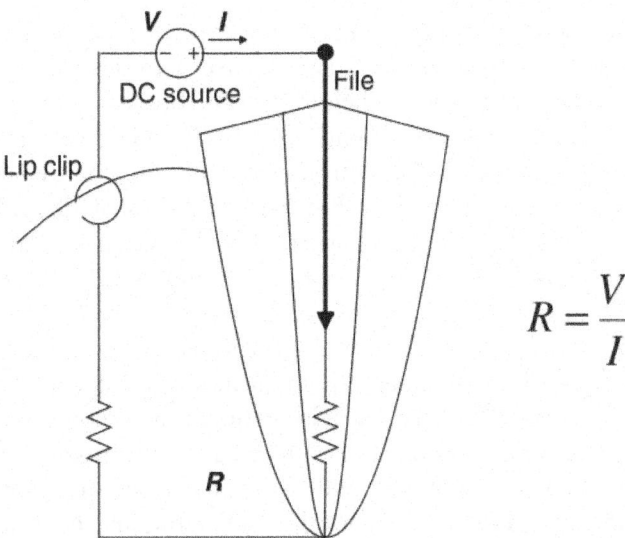

$$R = \frac{V}{I}$$

Figure 8: A simple resistive model of the apex, used in resistance-based ERCLMDs.

Although many of the resistance-based ERCLMDs have been shown to be accurate in dry conditions within the canal, it has been reported they were not always accurate when strong electrolytes, excessive haemorrhage, pus or pulp tissue was present (Suchde & Talim 1977, Nekoofar et al. 2002, Pommer et al. 2002, Tinaz et al. 2002). As soon as the file tip touches the electroconductive solution (electrolyte), in these situations, the DC voltage polarizes the tissues and varies its resistivity (Foster & Schwan 1989), thus completing the electrical circuit; the device incorrectly indicates that the minor apical foramen has been reached (Suchde & Talim 1977). Another disadvantage of the DC current is that an electric shock can be felt by the patient (Kim & Lee 2004). To eliminate the disadvantages of DC current Suchde & Talim (1977) proposed using AC current to measure the resistance. However, they still used a simple electronic ohmmeter, a 'bridge circuit', to overcome the disadvantages of the simple resistance method. The advantages of AC current are that it causes less damage to the tissue and improves functionality in 'wet' conditions as the resistivity of the electrolytes experience better stability (Suchde & Talim 1977, Foster & Schwan 1989). However, the disadvantage is that the capacitive component of the canal, which is variable with many parameters, will have an additional effect on the circuit. Therefore, in wet conditions when the capacitive component is more dominant, these devices suffer from a lack of accuracy (O'Neill 1974, Suchde & Talim 1977, Meredith & Gulabivala 1997).

Low frequency oscillation ERCLMDs

As discussed earlier, in practical endodontic experiments, the structure of the endodontic instrument, canal and tissues have capacitive characteristics as well as resistive characteristics. Therefore, modelling the circuit with a simple resistive characteristic is not sufficient (Inoue 1973, McDonald & Hovland 1990, Meredith & Gulabivala 1997). Unfortunately, these capacitive characteristic are variable and can change with the shape of the canal, and other physical parameters such as the dielectric constant of the liquids inside the canal. Based on this view, Inoue (1972, 1973) developed a different ERCLMD that worked through the principle of electrical resistance but was modified by the addition of an audible 'marker tone'. The principle of measuring root canal length by this device is based on the assumption that the low frequency oscillation, as produced by the resistance and capacity between the oral mucous membrane and the gingival sulcus, is the same as the frequency between the periodontal ligament (at the canal terminus) and the oral mucous membrane (Inoue & Skinner 1985). When the file reaches the canal terminus, the oscillating tones produced by the gingival sulcus and the canal terminus are coincident. As the impedance between the two electrodes is included in the feedback loop of the oscillator circuit, it influences the oscillated frequency value. Therefore, the frequency of the measured current changes as the load varies.

In other words, the impedance of the root canal depends on many parameters and is not the same in different canals. However, it can be assumed that the impedance between the oral mucous membrane and the depth of the gingival sulcus closely resembles the impedance between the canal terminus and the oral mucous membrane. Based on this assumption Inoue's device, the SonoExplorer (Hayashi Dental Supply, Tokyo, Japan), measures these two impedances and identifies the canal terminus when the readings approach each other. The frequency of this impedance is directed to a speaker that develops an auditory tone generated by means of low frequency oscillation.

The most important disadvantage of this device was the need for individual calibration. The device had to be calibrated at the periodontal sulcus in each tooth. The technique involved inserting a file with a silicon plastic-sheath into the gingival crevice of the tooth to be measured and the sound produced was named the 'gingival crevice sound'. Then, the conventional endodontic file was inserted into the root canal and when the sound produced by this file became 'identical' with the 'gingival crevice sound', the rubber stop was aligned with the reference point, and the measurement taken.

High Frequency Devices (Capacitancebased Devices) ERCLMDs

The Endocater (Hygenic Corp., Akron, OH, USA) was developed in 1979 by Hasegawa (Fouad & Krell 1989, Fouad et al. 1990, Pallares & Faus 1994) that used a high frequency (400 kHz) reference circuit (McDonald & Hovland 1990). To further decrease the influence of the variable capacitive characteristics on measurements insulated files were also used (Keller et al. 1991). As explained earlier (equation 3), the value of a capacitor is directly proportional to the area of its plates. Here, the insulator covers most of the surface of the file to decrease its capacitance value. Unfortunately, the coated file cannot be used in narrow canals, because the coating is easily abraded thus disturbing the measurement (Keller et al. 1991). In addition, Himel & Schott (1993) showed that the quality of the seal provided against electrical conductivity was decreased after autoclaving.

Capacitance and Resistance ERCLMDs (Look-up Table)

In 2003, the Elements™ Diagnostic Unit (SybronEndo, Anaheim, CA, USA) was introduced. It measured the capacitance and resistance of the circuit separately (Gordon & Chandler 2004, Vera & Gutierrez 2004). As explained earlier, these values are variable with many parameters. However, experimental look-up tables were developed that included the statistics of the values at different positions (Serota et al. 2004). The device exploits a composite signal with two frequencies to measure the resistance and capacitance of the system and then compares the measured values with its lookup table to diagnose the position of the file. As a result of modern electronic digital circuits, the manufacturer claims that this device has more consistent readings than its predecessors (Serota et al. 2004). Nevertheless, the fundamental principle of all the ERCLMDs is the same, i.e. an electrical model is assumed and the characteristics of that model are measured to diagnose the clinical properties. Based on clinical observations, Vera & Gutierrez (2004) reported that when using the Elements™ Diagnostic Unit the file should be withdrawn to the 0.5 mm mark instead of the 0.0 mm mark to achieve the accurate identification of the apical constriction that they assumed should be 0.5 mm short of the external (major) foramen. Therefore, taking the file to the 0.0 mark on the display and then withdrawing it 0.5 mm appears to be the most accurate way to use this device. In an attempt to achieve better results Vera & Gutierrez (2004) also recommended the access cavity should be dried before introducing the file into the canal.

Voltage Gradient ERCLMDs (Difference in Impedance with Three Nodes)

Several studies have been conducted to determine a constant electrical resistance or impedance in order to diagnose the position of the canal terminus, e.g. Sunada (1962) who determined R ¼ 6.5 k Ω. However, Meredith & Gulabivala (1997) reported there is neither constant reference impedance nor constant resistance for all root canals. The reason that, for example, resistance-based devices work in a reasonable number of cases is that there is a substantial difference between the resistance (or the impedance) value at the pulp and periodontal junction compared with intracanal positions. Indeed, this is the property Custer (1918) described many years ago.

Based on this fact, Ushiyama (1983) proposed a method to measure the variation in impedance when a file was inserted into the root canal. Using bipolar electrodes and applying a 400-Hz alternating current, this device monitored variations in the impedance value. Ushiyama (1983) concluded that a sharp variation in the value, determined the position of the file at the apical constriction, the narrowest portion of the root canal system. Ushiyama (1983) also reported that in the presence of strong electrolytes the 'voltage gradient method' could accurately detect the apical constriction. However, use of a special bipolar electrode is one of the main disadvantages of this device, as the electrode will not fit into narrow canals. No commercial product of the experimental device developed by Ushiyama (1983) appears to be available.

Two frequencies, Impedance difference ERCLMDs

Yamaoka (1984 – quoted in Saito & Yamashita 1990) developed a measurement device in which two frequencies were employed in the measurement. This device measures the impedance value at two different frequencies (f_H and f_L) and calculates the difference between the two values:

$$Diff = Z(f_H) - Z(f_L)$$

(7)

In fact, the actual measured signal is the difference between the voltages in two frequencies that is obviously proportional to the difference in impedance values. In the coronal portion of the root canal system, the device must be calibrated to eliminate any effect of the dielectric material inside the canal.

According to equation 3, the magnitude of the capacitance of the model is proportional to the distance between the two nodes shown in Fig. 6. That means, when the file approaches the canal terminus, the value of the capacitance sharply increases probably because of change in the morphology of the apical portion of the root. On the other hand, the f_H frequency used in this device is five times the f_L value. Therefore, according to equation 4, the change in $Z(f_L)$ will be five times larger than $Z(f_H)$, i.e. the difference between two $Z(f_L)$ and $Z(f_H)$ impedances rapidly increases at the 'apical foramen'. This method has been used in the Apit device (Osada, Tokyo, Japan).

According to Saito & Yamashita (1990) electrolytes, such as saline, 5% NaOCl, 14% EDTA and 3% H_2O_2 did not interfere with the detection of the apical terminus regardless of the size of the endodontic file or the size of 'apical foramen'. Frank & Torabinejad (1993) also confirmed that the location of the canal terminus could be detected under moist conditions, but due to the open electrical circuit the Apit cannot accurately detect the canal terminus in a dry canal. However, this phenomenon could be useful for checking the dryness of the root canal system prior to canal filling (Dahlin 1979).

Two Frequencies, Impedance Ratio (Quotient) ERCLMDs

In the impedance ratio-based ERCLMDs the AC source is again a two-frequency source, i.e. it comprises two sine waves with a high and a low frequency (f_H and f_L respectively). The impedance of the model is measured at each frequency and the position of the file is determined from the ratio between these two impedances:

$$Ratio = \frac{Z(f_H)}{Z(f_L)}$$

(8)

Kobayashi & Suda (1994) proved that the ratio had a definite value determined by the frequencies used and that the ratio indicates the location of the file tip in the canal. The quotient of the two impedances is nearly 1 when the tip of the file is some distance from the canal terminus. When the file is not at the minor apical foramen, the distance between the two plates of the model capacitance is high. Therefore, according to equation 3, the magnitude of the capacitance is negligible. Hence, the ratio of equation 8 will turn out to be a ratio of two equivalent resistance values which tends towards 1.

At positions close to the canal terminus, however, the capacitive characteristic of the impedance starts to appear. The influence of the capacitance on the overall impedance is proportional to the frequency of the measurement as shown in equation 4. At high frequencies (f_H) the overall impedance value will be much lower than at low frequency (f_L). That means, at the apical constriction the ratio tends towards a small value (Kobayashi & Suda 1994), however, this phenomenon is related to the morphology of the constriction. Lack of a constriction because of open apices (Hu"lsmann & Pieper 1989, Goldberg et al. 2002) or an impenetrable canal (Rivera & Seraji 1993, Ibarrola et al. 1999) have been reported as an impediment to determine the position of the canal terminus (Oishi et al. 2002). The ratio of equation 8 is independent of the electrolyte liquid inside the canal. This is because a change in the electrolyte material, which is a change in dielectric constant (e of equation 3), will influence equally the numerator and denominator of equation 8, and hence the final ratio will still remain constant. This concept underpins the development of the Root ZX (J. Morita Co., Kyoto, Japan), the first commercial ratiobased ERCLMD (Kobayashi 1995). This fundamental operating principle could explain why there was no statistically significant difference between their ability to determine the apical constriction in roots with vital pulps versus those with necrotic pulps (Dunlap et al. 1998) and/or various irrigants (Jenkins et al. 2001). Dunlap et al. (1998) reported that there was no statistical difference between the ability of the Root ZX to determine the apical constriction in vital canals versus necrotic canals.

Overall, the Root ZX was 82.3% accurate to within 0.5 mm of the apical constriction. In addition, this could also explain why this device was not adversely affected by the presence of sodium hypochlorite in the root canal system (Kobayashi 1995, Meares & Steiman 2002). Ounsi & Naaman (1999) in an ex vivo study reported that the Root ZX was not capable of detecting the apical constriction and should only be used to detect the major foramen. Hoer & Attin (2004) also showed that the use of impedance ratio electronic devices did not result in precise determination of the apical constriction, rather, under clinical conditions it was only possible to determine the region between the minor and major apical foramen. In their study, accurate determination of the apical constriction was only successful in 51–64.3% of canals, although the probability of determining the area between minor and major foramen was between 81 and 82.4% of cases. However, Shabahang et al. (1996) showed that when a potential error of ±0.5 mm from the 'foramen' was accepted as a clinically tolerable range, the Root ZX was able to locate the 'foramen' in 96.2% of cases.

Multifrequency ERCLMDs

There have been efforts to further increase the accuracy of ERCLMDs. One concept was to measure the impedance characteristics using more than two frequencies. In the Endo Analyzer 8005 (Analytic Endodontics, Sybron Dental, Orange, CA, USA) and AFA Apex Finder 7005 (Analytic Endodontics) five different frequencies have been used and the device measures both components (phase and amplitude) of impedance at each frequency. These figures are then analysed in a procedure to determine the location of the minor diameter (constriction) (Welk et al. 2003). The principle behind this device, however, is similar to the impedance ratio-based ERCLMDs. It detects the canal terminus by determining a sudden change in the dominant characteristic (capacitive or resistive) of the impedance. Welk et al. (2003) compared the accuracy of an impedance ratio-based ERCLMD (Root ZX) and the Endo Analyzer and found that the mean distance between the electronically located canal terminus and minor diameter was 1.03 mm for the Endo Analyzer and 0.19 mm for the Root ZX; the ability of the devices to locate the apical constriction was 34.4 and 90.7% of cases respectively. Pommer et al. (2002) evaluated the effect of pulp vitality on the accuracy of the AFA Apex Finder 7005 and reported that the difference between measurements in canals with vital or necrotic pulps was significantly different and concluded that the AFA Apex Finder was more accurate in vital cases.

SUMMARY

There is a general consensus that root canal procedures should be limited within the confines of the root canal, with the logical end-point for preparation and filling being the narrowest part of the canal. It is not possible to predictably detect the position of the apical constriction clinically, indeed, the constriction is not uniformly present, or may be irregular. Equally, it is not logical to base the end-point of root canal procedures on an arbitrary distance from the radiographic apex as the position of the apical foramen is not related to the 'apex' of the root.

Electronic root canal length measuring devices offer a means of locating the most appropriate end-point for root canal procedures, albeit indirectly.

The principle behind most ERCLMDs is that human tissues have certain characteristics that can be modelled by means of a combination of electrical components. Then, by measuring the electrical properties of the model (e.g. resistance, impedance) it should be possible to detect the canal terminus. Thus, most modern ERCLMDs are capable of recording the point where the tissues of the periodontal ligament begin outside the root canal, and hence from this a formula can be applied to ensure that preparation is confined within the canal.

Most reports suggest that 0.5 mm should be subtracted from the length of the file at the point when the device suggests that the file tip is in contact with the PDL (zero reading). This does not mean that the constriction is located; rather it means that the instrument is within the canal and close to the PDL. It is not appropriate to rely on any device reading 0.5 mm short of the foramen as this will often be inaccurate. The use of 'generation X' to describe and classify these devices is unhelpful, unscientific and perhaps best suited to marketing issues.

REFERENCES

1. Blaskovic-Subat V, Maricic B, Sutalo J (1992) Asymmetry of the root canal foramen. International Endodontic Journal 25, 158–64.

2. Bridges JE (2002) Non-perceptible body current ELF effects as defined by electric shock safety data. Bioelectromagnetics 23, 542–4.

3. Brunton P, Abdeen D, MacFarlane T (2002) The effect of an apex locator on exposure to radiation during endodontic therapy. Journal of Endodontics 28, 524–6.

4. Chugal N, Clive J, Spangberg L (2003) Endodontic infection: some biologic and treatment factors associated with outcome. Oral Surgery, Oral Medicine, Oral Pathology, Oral Radiology, and Endodontics 96, 81–90.

5. Coombs CF (1999) Electronic Instrument Handbook, 2nd edn. New York: McGraw-Hill Education.

6. Custer LE (1918) Exact methods of locating the apical foramen. Journal of the National Dental Association 5, 815–9.

7. Dahlin J (1979) Electrometric measuring of the apical foramen. A new method for diagnosis and endodontic therapy. Quintessence International 10, 13–22.

8. Dummer PMH, McGinn JH, Rees DG (1984) The position and topography of the apical canal constriction and apical foramen. International Endodontic Journal 17, 192–8.

9. Dunlap C, Remeikis N, BeGole E, Rauschenberger C (1998) An in vivo evaluation of an electronic apex locator that uses the ratio method in vital and necrotic canals. Journal of Endodontics 24, 48–50.

10. ElAyouti A, Weiger R, Lo"st C (2002) The ability of root ZX apex locator to reduce the frequency of overestimated radiographic working length. Journal of Endodontics 28, 116–9.

11. Forsberg J (1987a) A comparison of the paralleling and bisecting-angle radiographic techniques in endodontics. International Endodontic Journal 20, 177–82.

12. Forsberg J (1987b) Radiographic reproduction of endodontic 'working length' comparing the paralleling and the bisecting-angle techniques. Oral Surgery, Oral Medicine, and Oral Pathology 64, 353–60.

13. Foster K, Schwan H (1989) Dielectric properties of tissues and biological materials: a critical review. Critical Reviews in Biomedical Engineering 17, 25–104.

14. Fouad A, Krell K (1989) An in vitro comparison of five root canal length measuring instruments. Journal of Endodontics 15, 573–7.

15. Fouad A, Krell K, McKendry D, Koorbusch G, Olson R (1990) Clinical evaluation of five electronic root canal length measuring instruments. Journal of Endodontics 16, 446–9.

16. Frank AL, Torabinejad M (1993) An in vivo evaluation of Endex electronic apex locator. Journal of Endodontics 19, 177–9.

17. Gandhi OP (2002) Electromagnetic fields: human safety issues. Annual Review of Biomedical Engineering 4, 211–34.

18. Geddes LA, Baker LF (1967) The specific resistance of biological materials: a compendium of data for the biomedical engineer and physiologist. Medical & Biological Engineering 5, 271–93.

19. Goldberg F, De Silvio A, Manfre S, Nastri N (2002) In vitro measurement accuracy of an electronic apex locator in teeth with simulated apical root resorption. Journal of Endodontics 28, 461–3.

20. Gordon MPJ, Chandler NP (2004) Electronic apex locators. International Endodontic Journal 37, 1–13.

21. Green D (1956) A stereomicroscopic study of the root apices of 400 maxillary and mandibular anterior teeth. Oral Surgery, Oral Medicine, and Oral Pathology 9, 1224–32.

22. Green D (1960) Stereomicroscopic study of 700 root apices of maxillary and mandibular posterior teeth. Oral Surgery, Oral Medicine, and Oral Pathology 13, 728–33.

23. Hartov A, Mazzarese R, Reiss F et al. (2000) A multichannel continuously selectable multifrequency electrical impedance spectroscopy measurement system. IEEE Transactions on Biomedical Engineering 47, 49–58.

24. Heling B, Karmon A (1976) Determining tooth length with bisecting angle radiographs. Journal of the British Endodontic Society 9, 75–9.

25. Heling I, Gorfil C, Slutzky H, Kopolovic K, Zalkind M, SlutzkyGoldberg I (2002) Endodontic failure caused by inadequate restorative procedures: review and treatment recommendations. Journal of Prosthetic Dentistry 87, 674–8.

26. Himel V, Schott R (1993) An evaluation of the durability of apex locator insulated probes after autoclaving. Journal of Endodontics 19, 392–4.

27. Hoer D, Attin T (2004) The accuracy of electronic working length determination. International Endodontic Journal 37, 125–31.

28. Hu"lsmann M, Pieper K (1989) Use of an electronic apex locator in the treatment of teeth with incomplete root formation. Endodontics and Dental Traumatology 5, 238–41.

29. Ibarrola J, Chapman B, Howard J, Knowles K, Ludlow M (1999) Effect of preflaring on Root ZX apex locators. Journal of Endodontics 25, 625–6.

30. Inoue N (1972) Dental 'stethoscope' measures root canal. Dental Survey 48, 38–9.

31. Inoue N (1973) An audiometric method for determining the length of root canals. Journal of the Canadian Dental Association 39, 630–6.

32. Inoue N, Skinner DH (1985) A simple and accurate way of measuring root canal length. Journal of Endodontics 11, 421–7.

33. Jenkins J, Walker W, Schindler W, Flores C (2001) An in vitro evaluation of the accuracy of the root ZX in the presence of various irrigants. Journal of Endodontics 27, 209–11.

34. Katz A, Tamse A, Kaufman AY (1991) Tooth length determination: a review. Oral Surgery, Oral Medicine, and Oral Pathology 72, 238–42.

35. Keller M, Brown CJ, Newton C (1991) A clinical evaluation of the Endocater – an electronic apex locator. Journal of Endodontics 17, 271–4.

36. Kim E, Lee SJ (2004) Electronic apex locator. Dental Clinics of North America 48, 35–54.

37. Kobayashi C (1995) Electronic canal length measurement. Oral Surgery, Oral Medicine, and Oral Pathology 79, 226–31.

38. Kobayashi C, Suda H (1994) New electronic canal measuring device based on the ratio method. Journal of Endodontics 20, 111–4.

39. Kuttler Y (1955) Microscopic investigation of root apexes. Journal of the American Dental Association 50, 544–52.

40. Kuttler Y (1958) A precision and biologic root canal filling technique. Journal of the American Dental Association 56, 38–50.

41. Lazrak A, Griffin G, Gailey P (1997) Studying electric field effects on embryonic myocytes. Biotechniques 23, 736–41.

42. Lin L, Rosenberg P, Lin J (2005) Do procedural errors cause endodontic treatment failure? Journal of the American Dental Association 136, 187–93.

43. Malueg L, Wilcox L, Johnson W (1996) Examination of external apical root resorption with scanning electron microscopy. Oral Surgery, Oral Medicine, Oral Pathology, Oral Radiology, and Endodontics 82, 89–93.

44. Martinez-Lozano M, Forner-Navarro L, Sanchez-Cortes J, Llena-Puy C (2001) Methodological considerations in the determination of working length. International Endodontic Journal 34, 371–6.

45. McDonald NJ (1992) The electronic determination of working length. Dental Clinics of North America 36, 293–305.

46. McDonald NJ, Hovland EJ (1990) An evaluation of the Apex Locator Endocater. Journal of Endodontics 16, 5–8.

47. Meares W, Steiman H (2002) The influence of sodium hypochlorite irrigation on the accuracy of the Root ZX electronic apex locator. Journal of Endodontics 28, 595–8.

48. Meredith N, Gulabivala K (1997) Electrical impedance measurement of root canal length. Endodontics and Dental Traumatology 13, 126–31.

49. Morfis A, Sylaras S, Georgopoulou M, Kernani M, Prountzos F (1994) Study of the apices of human permanent teeth with the use of a scanning electron microscope. Oral Surgery, Oral Medicine, and Oral Pathology 77, 172–6.

50. Nekoofar MH (2005) Letter to the editor. International Endodontic Journal 38, 417–8.

51. Nekoofar MH, Sadeghi K, Sadaghi Akha E, Namazikhah MS (2002) The accuracy of the Neosono Ultima EZ apex locator using files of different alloys: an in vitro study. Journal of the California Dental Association 30, 681–4.

52. Niple J, Daigle J, Zaffanella L, Sullivan T, Kavet R (2004) A portable meter for measuring low frequency currents in the human body. Bioelectromagnetics 25, 369–73.

53. O'Neill L (1974) A clinical evaluation of electronic root canal measurement. Oral Surgery, Oral Medicine, and Oral Pathology 38, 469–73.

54. Oishi A, Yoshioka T, Kobayashi C, Suda H (2002) Electronic detection of root canal constrictions. Journal of Endodontics 28, 361–4.

55. Olson A, Goerig A, Cavataio R, Luciano J (1991) The ability of the radiograph to determine the location of the apical foramen. International Endodontic Journal 24, 28–35.

56. Ounsi H, Naaman A (1999) In vitro evaluation of the reliability of the Root ZX electronic apex locator. International Endodontic Journal 32, 120–3.

57. Pallares A, Faus V (1994) An in vivo comparative study of two apex locators. Journal of Endodontics 20, 576–9.

58. Palmer M, Weine F, Healey H (1971) Position of the apical foramen in relation to endodontic therapy. Journal of the Canadian Dental Association 37, 305–8.

59. Pendlebury ME, Horner K, Eaton KA (2004) Selection Criteria for Dental Radiography, 1st edn. London, UK: Faculty of General Dental Practitioners, Royal College of Surgeons of England, pp. 6–17.

60. Pommer O, Stamm O, Attin T (2002) Influence of the canal contents on the electrical associated determination of the length of root canals. Journal of Endodontics 2, 83–5.

61. Ponce EH, Fernandez JAV (2003) The cemento-dentino-canal junction, the apical foramen, and the apical constriction: evaluation by optical microscopy. Journal of Endodontics 29, 214–9.

62. Ricucci D (1998) Apical limit of root canal instrumentation and obturation, part 1. Literature review. International Endodontic Journal 31, 384–93.

63. Ricucci D, Langeland K (1998) Apical limit of root canal instrumentation and obturation, part 2. A histological study. International Endodontic Journal 31, 394–409.

64. Rivera E, Seraji M (1993) Effect of recapitulation on accuracy of electronically determined canal length. Oral Surgery, Oral Medicine, and Oral Pathology 76, 225–30.

65. Rush S, Abildskov JA, McFee R (1963) Resistivity of body tissues at low frequencies. Circulation Research 12, 40–50.

66. Saito T, Yamashita Y (1990) Electronic determination of root canal length by newly developed measuring device – influence of the diameter of apical foramen, the size of K-file and the root canal irrigants. Dentistry in Japan 27, 65–72.

67. Schaeffer M, White R, Walton R (2005) Determining the optimal obturation length: a meta-analysis of the literature. Journal of Endodontics 31, 271–4.

68. Seltzer S (1988) Endodontology, 2nd edn. Philadelphia: Lea & Febriger, pp. 24–30.

69. Serota KS, Vera J, Barnett F, Nahmias Y (2004) The new era of foramenal location. Endodontic Practice 7, 17–22.

70. Shabahang S, Goon W, Gluskin A (1996) An in vivo evaluation of Root ZX electronic apex locator. Journal of Endodontics 22, 616–8.

71. Sheaffer J, Eleazer P, Scheetz J, Clark S, Farman A (2003) Endodontic measurement accuracy and perceived radiograph quality: effects of film speed and density. Oral Surgery, Oral Medicine, Oral Pathology, Oral Radiology, and Endodontics 96, 441–8.

72. Stein T, Corcoran J (1990) Anatomy of the root apex and its histologic changes with age. Oral Surgery, Oral Medicine, and Oral Pathology 69, 238–42.

73. Suchde RV, Talim SD (1977) Electronic ohmmeter: an electronic device for the determination of the root canal length. Oral Surgery, Oral Medicine, and Oral Pathology 43, 141–9.

74. Sunada I (1962) New method for measuring the length of the root canal. Journal of Dental Research 41, 375–87.

75. Suzuki K (1942) Experimental study on iontophoresis. Japanese Journal of Stomatology 16, 411–29.

76. Tinaz AC, Sibel Sevimli L, Gorgul G, Turkoz EG (2002) The effects of sodium hypochlorite concentrations on the accuracy of an apex locating device. Journal of Endodontics 28, 160–2.

77. Trope M, Rabie G, Tronstad L (1985) Accuracy of an electronic apex locator under controlled clinical conditions. Endodontics and Dental Traumatology 1, 142–5.

78. Ushiyama J (1983) New principle and method for measuring the root canal length. Journal of Endodontics 9, 97–104.

79. Van Driessche W, De Vos R, Jans D, Simaels J, De Smet P, Raskin G (1999) Transepithelial capacitance decrease reveals closure of lateral

interspace in A6 epithelia. Pflu¨gers Archives European Journal of Physiology 437, 680–90.

80. Vera J, Gutierrez M (2004) Accurate working-length determination using a Fourth-generation apex locator. Contemporary Endodontics 1, 4–8.

81. Vertucci F (2005) Root canal morphology and its relationship to endodontic procedures. Endodontic Topics 10, 3–29.

82. Welk AR, Baumgartner JC, Marshall JG (2003) An in vivo comparison of two frequency-based electronic apex locators. Journal of Endodontics 29, 497–500.

CITATION

CHAPTER 1

El-Hassane Aglzim, Amar Rouane and Reddad El-Moznine, An Electronic Measurement Instrumentation of the Impedance of a Loaded Fuel Cell or Battery, doi:10.3390/s7102363

CHAPTER 2

Prasad Calyam, Abdul Kalash, Ramya Gopalan, Sowmya Gopalan, and Ashok Krishnamurthy, "RICE: A Reliable and Efficient Remote Instrumentation Collaboration Environment," Advances in Multimedia, vol. 2008, Article ID 615186, 17 pages, 2008. doi:10.1155/2008/615186

CHAPTER 3

Lorena Arruda et al. Efficacy of Electronic Foramen Locators in Controlling Root Canal Working Length during Rotary Instrumentation. *Braz. Dent. J.* [online]. 2015, vol.26, n.5 [cited 2016-03-10], pp. 547-551 . Available from: <http://www.scielo.br/scielo.php?script=sci_arttext&pid=S0103-64402015000500547&lng=en&nrm=iso>. ISSN 1806-4760. http://dx.doi.org/10.1590/0103-6440201300099.

CHAPTER 4

Zarate JM, Ritson CR, Poeppel D (2013) The Effect of Instrumental Timbre on Interval Discrimination. PLoS ONE 8(9): e75410. doi:10.1371/journal.pone.0075410

CHAPTER 5

D. Fisher and P. Gould, "Open-Source Hardware Is a Low-Cost Alternative for Scientific Instrumentation and Research," Modern Instrumentation, Vol. 1 No. 2, 2012, pp. 8-20. doi: 10.4236/mi.2012.12002.

CHAPTER 6

1Gilsa Aparecida de Lima Machado, Patricia Mara Danella Zacaro, Alderico Rodrigues de Paula Junior and Marcelo Lopes de Oliveira e Souza Research and Development Institute (IPD), Applying Software Engineering Methodology for Designing Biomedical Software Devoted to Electronic Instrumentation, ISSN 1549-3636

CHAPTER 7

Chi-Kuen Lo (2013). Instrumentation for Ferromagnetic Resonance Spectrometer, Ferromagnetic Resonance - Theory and Applications, Dr. Orhan Yalçın (Ed.), ISBN: 978-953-51-1186-3, InTech, DOI: 10.5772/56069.

CHAPTER 8

Li-Te Yin, Hung-Yu Wang, Yang-Chiuan Lin and Wen-Chung Huang, A Novel Instrumentation Circuit for Electrochemical Measurements, doi:10.3390/s120709687.

CHAPTER 9

Özgür Genç, Tayfun Alaçam, Guven Kayaoglu, Evaluation of three instrumentation techniques at the precision of apical stop and apical sealing of obturation, http://dx.doi.org/10.1590/S1678-77572011005000009.

CHAPTER 10

Gaetano D. Gargiulo, True Unipolar ECG Machine for Wilson Central Terminal Measurements, http://dx.doi.org/10.1155/2015/586397.

CHAPTER 11

Ali Bulent Usakli, Improvement of EEG Signal Acquisition: An Electrical Aspect for State of the Art of Front End, doi:10.1155/2010/630649.

CHAPTER 12

Filatov, A. (2015) Application Concept of Zero Method Measurement in Microwave Radiometers. Modern Instrumentation, 4, 19-31. doi: 10.4236/mi.2015.43003.

CHAPTER 13

Maghsoudi, A. and Kalantari, B. (2014) Monitoring Instrumentation in Underground Structures. *Open Journal of Civil Engineering*, 4, 135-146. doi: 10.4236/ojce.2014.42012.

CHAPTER 14

M. H. Nekoofar, M. M. Ghandi , S. J. Hayes & P. M. H. Dummer, The fundamental operating principles of electronic root canal length measurement devices, doi:10.1111/ j.1365-2591.2006.01131.x.

INDEX

A

Alternating current (AC) 260
Analog digital converter (ADC) 208
Analog-to-digital (A/D) 88
Analog to digital converter (ADC) 159, 211
Apical constriction 172
Auto-reverse mode (AAR) 55

B

Brain Computer Interface (BCI) 203

C

Cemento-dentinal junction (CDJ) 253
Center for Accelerated Maturation of Materials (CAMM) 21
Common mode rejection ratio (CMRR) 208
Common Mode Rejection Ratio (CMRR) 186
Computerized Electronic Instrumentation (CEI) 112
Cone penetration Tests (CPT) 237
Co-planar waveguide (CPW) 144

D

Data Flow Diagram (DFD) 114, 121, 128
Device under test (DUT) 6
Direct current (DC) 260
Directional coupler (DC) 226

E

Electrical Source Imaging (ESI) 203
Electrochemical impedance spectroscopy (EIS) 2
Electrochemical Impedance Spectroscopy (EIS) 4
Electroencephalogram (EEG) 199
Electromagnetic interference (EMI) 207
Electronic foramen locators (EFLs) 55, 56
Electronic root canal length measurement device' (ERCLMD) 255
Electrostatic discharge (ESD) 208
Endodontic treatment 172
Energy dispersive spectroscopy (EDS) 40
Entity and Relationship Diagram (ERD) 118

Event detected (ED) 44
Excavation Damage Zone (EDZ) 246
Extended gate field effect transistor
(EGFET) 158

F

False alarm (FA) 70
Fluid filtration 171, 175, 178

G

Geotechnical construction 235, 236
Gigabyte (GB) 96
Graphical user interface (GUI) 25

I

Infrared thermometer (IRT) 91
Input/output (IO) 88
Instrumentation amplifier (IA) 159
Integrated Development Environment
(IDE) 87
International Standard Organization
(ISO) 114
Ion-sensitive field effect transistor (IS-FET) 158

K

Kilobytes (KB) 88

L

Laboratory of Electronic Instrumentation
of Nancy (L.I.E.N) 2
Left arm (LA) 184
Left leg (LL) 184
Light-emitting diodes (LEDs) 106
Limit-of-detection (LOD) 167
Limit-of-quantification (LOQ) 167
Liquid crystal display module (LCM)
159

M

Manual instrumentation 173, 178
Modulation amplitude (MA) 148
Modulation frequency (MF) 148

N

National Center for Microscopy and
Imaging Research (NCMIR) 23
New York University (NYU) 68
Noise generator (NG) 225

O

Ohio Supercomputer Center's (OSC) 24

P

Phosphate buffer (PB) 163
Physiologic saline solution 173
Pressure meter tests (PMT) 237
Printed circuit board (PCB) 105

Q

Quality of experience (QoE) 19, 20

R

Real-time clock (RTC) 95
Reference electrode (RE) 159
Remote Instrumentation Collaboration
Environment (RICE) 19, 22, 37
Right arm (RA) 184
Right leg (RL) 184
Round-trip delay (RTT) 38

S

Scalar network analyzer (SNA) 150
Seismic refraction tomography (SRT)
245
Shorted waveguide cavity (SWC) 143
Signal to noise ratio (SNR) 138, 147
Signal to Noise Ratio (SNR) 209
Software Engineering (SE) 111, 112
Standard Penetration Test (SPT) 237
Stereomicroscope 174

T

Time constant (TC) 148
Trigger elevated (TE) 44

U

Ultra High Voltage Electron Microscopy (UHVEM) 23
Unified Modelling Language (UML) 118

V

Vector network analyzer (VNA) 137
Virtual network computing (VNC) 22

W

Wilson Central Terminal (WCT) 186
Working electrode (WE) 159